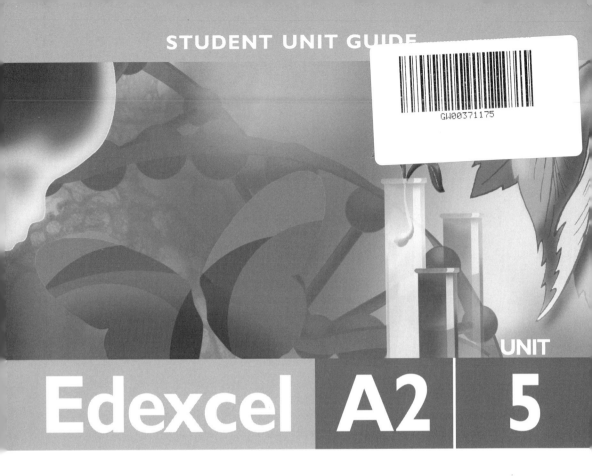

UNIT

Edexcel A2 5

Biology

Energy, Exercise and Coordination

Mary Jones

Philip Allan Updates, an imprint of Hodder Education, an Hachette UK company, Market Place, Deddington, Oxfordshire OX15 0SE

Orders

Bookpoint Ltd, 130 Milton Park, Abingdon, Oxfordshire OX14 4SB
tel: 01235 827720
fax: 01235 400454
e-mail: uk.orders@bookpoint.co.uk

Lines are open 9.00 a.m.–5.00 p.m., Monday to Saturday, with a 24-hour message answering service. You can also order through the Philip Allan Updates website: www.philipallan.co.uk

© Philip Allan Updates 2009

ISBN 978-0-340-94832-3

First printed 2009
Impression number 5 4 3 2 1
Year 2014 2013 2012 2011 2010 2009

This guide has been written specifically to support students preparing for the Edexcel A2 Biology Unit 5 examination. The content has been neither approved nor endorsed by Edexcel and remains the sole responsibility of the author.

Typeset by Greenhill Wood Studios
Printed by MPG Books, Bodmin

Hachette UK's policy is to use papers that are natural, renewable and recyclable products and made from wood grown in sustainable forests. The logging and manufacturing processes are expected to conform to the environmental regulations of the country of origin.

Contents

Introduction

About this guide 4

The specification 4

Scientific language 7

Revision 7

The examination 8

■ ■ ■

Content Guidance

About this section 12

Run for your life 13

Grey matter 29

■ ■ ■

Questions and Answers

About this section 46

Scientific article 1 47

Scientific article 2 55

Sample paper 1 61

Sample paper 2 80

Introduction
About this guide

This book is the fourth in a series of four, which will help you to prepare for the Edexcel A-level biology examination. It covers **Unit 5: Energy, Exercise and Coordination**. This is the second of two content-based units that make up the A2 biology examination. The other three books in the series cover Units 1, 2 and 4.

This guide has three main sections:
- **Introduction** This contains an overview of the unit and how it is assessed, some advice on revision and advice on doing the examination.
- **Content Guidance** This provides a summary of the facts and concepts that you need to know for the Unit 5 examination.
- **Questions and Answers** This section contains two specimen papers for you to try, each worth 90 marks. There are also two sets of answers for each question, one from a candidate who is likely to get a C grade and another from a candidate who is likely to get an A grade.

It's entirely up to you how you use this book. We suggest you start by reading through this Introduction, which will give you some suggestions about how you can improve your knowledge and skills in biology and about good ways of revising. It also gives you some pointers into how to do well in the examination. The Content Guidance will be especially useful when you are revising, as will the Questions and Answers.

The specification

It is a good idea to have your own copy of the Edexcel biology specification. It's you who is going to take this examination, not your teacher, and so it is your responsibility to make sure you know as much about the exam as possible. You can download a copy free from **www.edexcel.org.uk**.

The A2 examination is made up of three units:
- Unit 4 The Natural Environment and Species Survival
- Unit 5 Energy, Exercise and Coordination
- Unit 6 Practical Biology and Investigative Skills

This book covers Unit 5. There is no book for Unit 6, because this is based on practical work that you will do in your biology classes.

Concept-led or context-led?

There are two ways of working through the specification, and your teacher will be able to tell you which one you are following. He or she may be taking a context-led approach, in which you start with a context such as a person having a heart attack, and then look at biological facts, principles and concepts associated with this. Or you may be following a concept-led approach, in which you start with the biological facts, principles and concepts and then look at how these can be applied to particular contexts.

In the end, there is not much difference — you need to learn all the biology whichever way you approach it! In this book we have followed a concept-led approach, but you'll find it easy to follow, whichever approach you have been taking in your biology course.

Unit 5 content

The content of each unit is clearly set out in the specification. Unit 5 has two topics:
- Run for your life
- Grey matter

Run for your life begins by looking at the fine structure of muscles, and how they contract, as well as their interaction with the skeleton. The energy for muscle contraction comes from respiration, and both aerobic and anaerobic respiration are explained. You need to know how to investigate the rate of respiration practically, for example using a respirometer.

We then look at the function of cardiac muscle in the heart, and how all the different parts of the heart are coordinated in their activity by electrical impulses arising in the sinoatrial node. During exercise, heart rate and ventilation rate both increase, and we consider the various systems that are involved in their control. Exercise also causes an increase in body temperature, and this is regulated by the hypothalamus and various effectors. Long-term regulation involves the hormone thyroxine, and we look briefly at how this interacts with transcription factors to affect gene expression in cells.

Exercise is generally considered to be good for your health, and it does indeed help to reduce obesity, the risk of developing type II diabetes or coronary heart disease. However, too much exercise can be harmful, and we look at how it can lead to damaged joints or to reduced effectiveness of the immune system. You also need to discuss the pros and cons of banning the use of performance-enhancing drugs in sport.

Grey matter considers the structure and function of the various components of the human nervous system. The transmission of nerve impulses in the form of action potentials is described, and also the way in which these impulses are transmitted across synapses. We look in detail at one example of a receptor — the eye — and how rods can cause the generation of action potentials that transmit information about our environment to the brain.

Plants have coordination systems, too, and we look at these briefly and compare them with those of animals.

People have always been fascinated by the functions of the different parts of the brain, and we now have powerful tools to help us to determine these, such as fMRI and CT scans. We take a brief look at how one particular ability develops in the brain, and how researchers have used animal models to try to develop explanations for how our own brains develop and function, including how we learn. The 'nature vs nurture' debate is considered in relation to brain development in humans and other animals.

There are numerous serious illnesses, such as Parkinson's disease, that involve the brain, and research continues to find drugs that may help to alleviate their symptoms. We look at the causes of some of these illnesses and the drugs that can be used to treat them. The results of the Human Genome Project are giving us many new avenues to explore, including the possible development of drugs to suit individuals with particular genotypes. We consider the use of animals in this kind of research, and also the potential use of genetically modified organisms.

Unit 5 assessment

Unit 5 is assessed in an examination lasting 1 hour 30 minutes. The questions are all structured — that is, they are broken up into several parts, with spaces in which you write your answers. There are 90 marks available on the paper.

What is assessed?

It's easy to forget that your examination isn't just testing what you *know* about biology — it's also testing your *skills*. It's difficult to overemphasise how important these are.

The Edexcel examination tests three different assessment objectives (AOs). The following table gives a breakdown of the proportion of marks awarded to knowledge and to skills in the A2 examination:

Assessment objective	Outline of what is tested	Percentage of marks
AO1	Knowledge and understanding of science and of How Science Works	26–30
AO2	Applications of knowledge and understanding of science and of How Science Works	42–48
AO3	How Science Works	26

AO1 is about remembering and understanding all the biological facts and concepts you have covered in this unit. AO2 is about being able to *use* these facts and concepts in new situations. The examination paper will include questions that contain unfamiliar contexts or sets of data, which you will need to interpret in the light of the biological knowledge you have. When you are revising, it is important that you try to develop your ability to do this, as well as learning the facts.

AO3 is about How Science Works. Note that this comes into AO1 and AO2 as well. A science subject such as biology is not just a body of knowledge. Scientists do research to find out how things around them work, and new research continues to find out new things all the time. Sometimes new research means that we have to change our ideas. For example, not all that long ago people were encouraged to eat a lot of eggs and drink a lot of milk, because it was thought to be 'healthy'. Now we know we need to take care not to eat too many animal-based fats, because new research has found links between a fatty diet and heart disease.

How Science Works is about developing theories and models in biology, and testing them. It involves doing experiments to test hypotheses, and analysing the results to determine whether the hypothesis is supported or disproved. You need to appreciate why science does not always give us clear answers to the questions we ask, and how we can design good experiments whose results we can trust.

Scientific language

Throughout your biology course, and especially in your examination, it is important to use clear and correct biological language. Scientists take great care to use language precisely. If doctors or researchers do not use exactly the correct word when communicating with someone, then what they are saying could easily be misinterpreted. Biology has a huge number of specialist terms (probably more than any other subject you can choose to study at A2) and it is important that you learn them and use them. Your everyday conversational language, or what you read in the newspaper or hear on the radio, is often not the kind of language required in a biology examination. Be precise and careful in what you write, so that an examiner cannot possibly misunderstand you.

Revision

There are many different ways of revising, and what works well for you may not be as suitable for someone else. Have a look at the suggestions below and try some of them out.

- **Revise continuously.** Don't think that revision is something you do just before the exam. Life is much easier if you keep revision ticking along all through your biology course. Find 15 minutes a day to look back over work you did a few weeks ago, to keep it fresh in your mind. You will find this helpful when you come to start your intensive revision.
- **Understand it.** Research has shown that we learn things much more easily if our brain recognises that they are important to us and that they make sense to us.

Before you try to learn a topic, make sure that you understand it. If you don't, ask a friend or a teacher, find a different textbook in which to read about it, or look it up on the internet. Work at it until you feel you have got it sorted and then try to learn it.

- **Make your revision active.** Just reading your notes or a textbook won't do any harm, but it won't do all that much good, either. Your brain only puts things into its long-term memory if it thinks they are important, so you need to convince it that they are. You can do this by making your brain *do* something with what you are trying to learn. So, if you are revising a table comparing the structure of prokaryotic and eukaryotic cells, try rewriting it as bullet points. If you are revising independent assortment from a flow diagram, try rewriting it as a paragraph of text. Some people like drawing spider diagrams. You will learn much more by constructing your own table, flow diagram or set of bullet points than just trying to remember one that someone else has constructed.

- **Fair shares for all.** Don't always start your revision in the same place. If you always start at the beginning of the unit, then you will learn a lot about photosynthesis but not much about immunity. Make sure each part of the specification gets its fair share of your attention.

- **Plan your time.** You may find it helpful to draw up a revision planner, setting out what you will revise and when. Even if you don't stick to it, it will give you a framework that you can refer to — if you get behind with it, you can rewrite the next bits of the plan to squeeze in the topics you haven't yet covered.

- **Keep your concentration.** It's often said that it is best to revise in short periods, say 20 minutes or half an hour. This is true for many people, if they find it difficult to concentrate for longer than that. But there are others who actually find it better to settle down for a much longer period of time — even several hours — and really get into their work and stay concentrated without interruptions. Find out which works best for you. It may be different at different times of day. Maybe you can only concentrate well for 30 minutes in the morning, but are able to get lost in your work for several hours in the evening.

- **Don't assume you know it.** The topics where exam candidates are least likely to do well are, strangely, the ones that they have already learned something about at GCSE or AS. This is probably because if you think you already know something then you give that a low priority when you are revising. It's important to remember that what you knew for GCSE or AS is probably not detailed enough for A2.

The examination

Once you are in the examination room, at least you can stop worrying about whether or not you've done enough revision. The important thing now is to make the best use of the knowledge and understanding and skills that you have managed to get into your brain.

Time

You will have 90 minutes to answer questions worth 90 marks. That gives you 1 minute per mark. When you are trying out a test question, time yourself. Are you working too fast? Or are you taking too long? Get used to what it feels like to work at just over a-mark-a-minute rate.

It's not a bad idea to spend one of those minutes just skimming through the exam paper before you start writing. Maybe one of the questions looks as though it is going to need a bit more of your time than the others. If so, make sure you leave a little bit of extra time for it.

Read the question carefully

That sounds obvious, but candidates lose large numbers of marks by not doing it.
- There is often vital information at the start of the question that you'll need in order to answer the questions themselves. Don't just jump straight to the first place where there are answer lines and start writing. Start reading at the beginning! Examiners are usually careful not to give you unnecessary information, so if it is there it is probably needed. You may like to use a highlighter to pick out any particularly important bits of information in the question.
- Do look carefully at the command words (the ones right at the start of the question) and do what they say. For example, if you are asked to *explain* something then you won't get many marks — perhaps none at all — if you *describe* it instead. You can find all these words in an appendix near the end of the specification document.

Depth and length of answer

The examiners will give you two useful guidelines about how much you need to write.
- **The number of marks.** Obviously, the more marks the more information you need to give. If there are 2 marks, then you'll need to give two different pieces of information in order to get both of them. If there are 5 marks, you'll need to write much more.
- **The number of lines.** This isn't such a useful guideline as the number of marks, but it can still help you to know how much to write. If you find your answer won't fit on the lines, then you probably haven't focused sharply enough on the question. The best answers are short and precise.

Writing, spelling and grammar

The examiners are testing your biology knowledge and skills, not your English skills. Still, if they can't understand what you have written then they can't give you any marks. It is your responsibility to communicate clearly — don't scribble so fast that the examiner cannot read what you have written.

In general, incorrect spellings are not penalised. If the examiner knows what you are trying to say then he or she will give you credit. However, if your wrongly spelt word could be confused with another, then you won't be given the mark. For example, if you write 'meitosis', then the examiner can't know whether you mean meiosis or mitosis, so you'll be marked wrong.

Like spelling, bad grammar isn't taken into account. Once again, though, if it is so bad that the examiner cannot understand you, then you won't get marks. A common problem is to use the word 'it' in such as way that the examiner can't be certain what 'it' refers to. A good general rule is never to use this word in an exam answer.

Content
Guidance

This section of the guide summarises what you need to know for the Unit 5 test. It is divided into two topics:

(1) Run for your life
- Structure and function of muscles
- Respiration
- Function of the heart
- Physiological effects of exercise
- Temperature regulation
- Exercise and health
- Performance-enhancing drugs in sport

(2) Grey matter
- Photoreceptors in plants
- The mammalian nervous system
- The human brain
- Development of visual capacity
- Drugs and the brain

Run for your life

Muscles

Muscle tissue is made up of highly specialised cells that are able to use energy from the hydrolysis of ATP to make themselves shorter. This is called **contraction**.

The type of muscle attached to the bones of the skeleton is called striated (striped) muscle, because of its appearance when seen using a microscope. The 'cells' in striated muscle are highly specialised, and are called **muscle fibres**. Each fibre contains many nuclei, many mitochondria, a very extensive endoplasmic reticulum (sarcoplasmic reticulum) with infoldings called T-tubules, and fibrils made of two proteins — **actin** and **myosin**.

A short length of muscle fibre

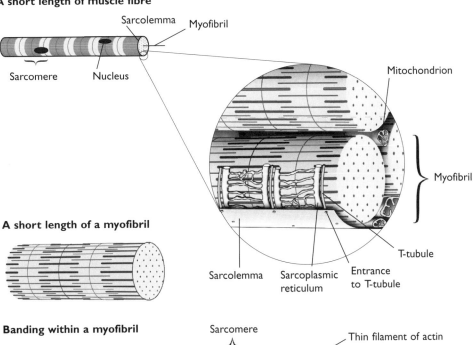

Sarcolemma Myofibril

Sarcomere Nucleus

Mitochondrion

Myofibril

A short length of a myofibril

T-tubule

Sarcolemma Sarcoplasmic reticulum Entrance to T-tubule

Banding within a myofibril Sarcomere

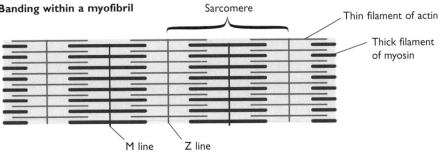

Thin filament of actin

Thick filament of myosin

M line Z line

The structure of a muscle fibre

The sliding filament theory

Muscles contract as the actin and myosin filaments slide between each other. This makes each sarcomere shorter, so the whole muscle fibre gets shorter.

- An action potential arrives at the cell surface membrane (sarcolemma) of a muscle fibre. It travels along the membranes of the sarcoplasmic reticulum, deep into the muscle fibre. This causes calcium ions (Ca^{2+}), which were stored in the sarcoplasmic reticulum, to be released into the muscle fibre.
- In a relaxed myofibril, proteins called **tropomyosin** and **troponin** cover binding sites on the actin filaments. The calcium ions cause the tropomyosin and troponin to change shape, which uncovers the binding sites.
- Heads of myosin molecules next to the uncovered binding sites now bind with the actin filaments, forming 'bridges' between them.
- The myosin heads tilt, pushing the actin filaments along.
- ATP now binds with the myosin heads and is hydrolysed by ATPase, releasing energy. The energy causes the myosin heads to disconnect from the actin filaments.
- The disconnected myosin heads flip back to their original position, and bind with another exposed binding site on the actin filament. They tilt again, pushing the actin filament along again.
- This process repeats over and over, as long as action potentials keep arriving.

The diagram shows how this process causes the actin and myosin filaments to slide between one another, and how this causes the muscle fibre to shorten.

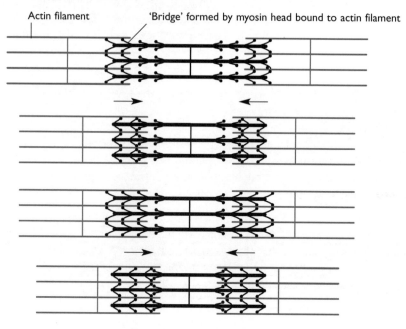

How sliding filaments cause muscle contraction

Fast twitch and slow twitch muscle fibres

We have two types of muscle fibres in our striated muscle tissue. Fast twitch fibres are adapted for rapid contraction over a short time period. Slow twitch fibres are adapted for slightly less rapid contraction over longer time periods.

Slow twitch fibres	Fast twitch fibres
Use aerobic respiration to produce ATP	Use anaerobic respiration to produce ATP
Have many mitochondria to carry out Krebs cycle and oxidative phosphorylation	Have few mitochondria
Have large amounts of the protein myoglobin, which stores oxygen	Have much less myoglobin
Are relatively narrow, so oxygen can diffuse into their centres rapidly	Are relatively wide
Are supplied with oxygenated blood by many capillaries	Few capillaries supplying them

Muscles and the skeleton

Striated muscles are attached to the skeleton by strong, inelastic cords called **tendons**. These are made of long fibres of the protein **collagen**, together with small amounts of another protein, **elastin**. When the muscle contracts, it pulls on the tendons which transmit the force to a bone.

Bones are connected at joints. Joints where the bones can move with respect to each other, such as the finger joints or elbow joint, are **synovial** joints. The bones at synovial joints are held together by **ligaments**. These, like tendons, contain collagen and elastin, but with a much greater proportion of elastin, which means that they can stretch much more than tendons.

Muscles can only produce a force when they contract. When they relax, they stay in the same position unless something pulls them back to their original, lengthened state. At a joint such as the elbow joint, some muscles pull in one direction when they contract, and others pull in the opposite direction. The major muscle causing the arm to bend at the elbow is the biceps, and this is called a **flexor** muscle. The major muscle causing the arm to straighten when it contracts is the triceps, and this is an **extensor** muscle.

When the arm bends, the biceps muscle contracts and the triceps generally relaxes, although it may contract a little to make sure the movement is controlled and steady. To straighten the arm, the triceps contracts and the biceps generally relaxes, although again it may contract enough to control the movement. These two muscles are said to be an **antagonistic muscle pair**.

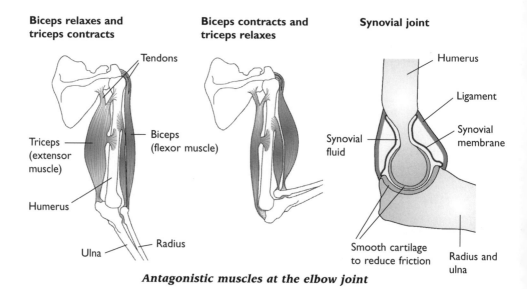

Antagonistic muscles at the elbow joint

Respiration

All cells obtain useable energy through respiration. Respiration is the oxidation of energy-containing organic molecules, such as glucose. These are known as **respiratory substrates**. The energy released from this process is used to combine ADP with inorganic phosphate to make ATP. ATP is the energy currency of cells (see page 13 in the *Unit 4 Student Unit Guide*).

In aerobic respiration, oxygen is involved. Glucose, $C_6H_{12}O_6$, (or another respiratory substrate) is split to release carbon dioxide as a waste product. The hydrogen from the glucose is combined with atmospheric oxygen. This releases a large amount of energy, which is used to drive the synthesis of ATP.

Measuring the rate of respiration

You can measure the rate of uptake of oxygen using a respirometer, shown in the diagram.

The organisms to be investigated are placed in one tube, and non-living material of the same mass in the other tube. Soda lime is placed in each tube, to absorb all carbon dioxide. Cotton wool prevents contact of the soda lime with the organisms.

Coloured fluid is poured into the reservoir of each manometer and allowed to flow into the capillary tube. It is essential that there are no air bubbles. You must end up with exactly the same quantity of fluid in the two manometers.

Two rubber bungs are now taken, fitted with tubes as shown in the diagram. Close the spring clips. Attach the manometers to the bent glass tubing, ensuring a totally airtight connection. Next, place the bungs into the tops of the tubes.

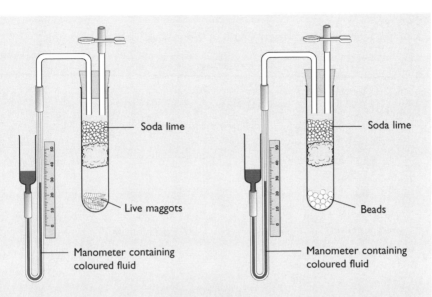

A respirometer

Open the spring clips. (This allows the pressure throughout the apparatus to equilibrate with atmospheric pressure.) Note the level of the manometer fluid in each tube. Close the clips. Each minute, record the level of the fluid in each tube.

As the organisms respire, they take oxygen from the air around them and give out carbon dioxide. The removal of oxygen from the air inside the tube reduces the volume and pressure, causing the manometer fluid to move towards the organisms. The carbon dioxide given out is absorbed by the soda lime (otherwise, the volume of carbon dioxide given out would match the volume of oxygen taken in, so the manometer fluid would not move).

You would not expect the manometer fluid in the tube with no organisms to move, but it may do so because of temperature changes. This allows you to control for this variable, by subtracting the distance moved by the fluid in the control manometer from the distance moved in the experimental manometer (connected to the living organisms), to give you an adjusted distance moved.

Calculate the mean (adjusted) distance moved by the manometer fluid per minute. If you know the diameter of the capillary tube, you can convert the distance moved to a volume:

volume of liquid in a tube = length $\times \pi r^2$

This gives you a value for the volume of oxygen absorbed by the organisms per minute.

You can compare rates of respiration at different temperatures by standing the apparatus in a water bath.

Glycolysis

Glycolysis is the first stage of respiration. It takes place in the cytoplasm.

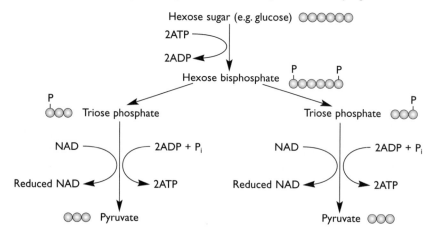

Glycolysis

- A glucose (or other hexose sugar) molecule is phosphorylated, as two ATPs donate phosphate to it.
- This produces a hexose bisphosphate molecule, which splits into two triose phosphates.
- Each triose phosphate is converted to a **pyruvate** molecule. (Pyruvate may also be referred to as pyruvic acid.) This involves the removal of hydrogens, which are taken up by a coenzyme called **NAD**. This produces reduced NAD. During this step, the phosphate groups from the triose phosphates are added to ADP to make ATP.
- Overall, two molecules of ATP are used and four are made during glycolysis of one glucose molecule, making a net gain of two ATPs per glucose.

The link reaction

If oxygen is available, the pyruvate now moves into a mitochondrion, where the link reaction and the Krebs cycle take place. During these processes, the glucose is completely oxidised.

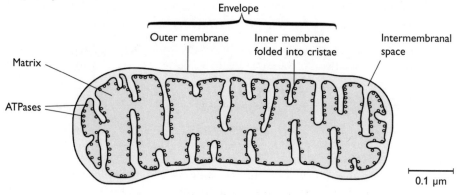

A mitochondrion

content guidance

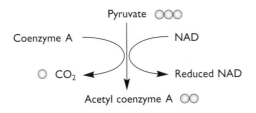

The link reaction

Carbon dioxide is removed from the pyruvate. This carbon dioxide diffuses out of the mitochondrion and out of the cell. Hydrogen is also removed from the pyruvate, and is picked up by NAD, producing reduced NAD. This converts pyruvate into a two-carbon compound. This immediately combines with coenzyme A to produce acetyl coenzyme A.

The Krebs cycle

Acetyl coenzyme A has two carbon atoms. It combines with a four-carbon compound to produce a six-carbon compound. This is gradually converted to the four-carbon compound again through a series of enzyme-controlled steps. These steps all take place in the matrix of the mitochondrion, and each is controlled by specific enzymes.

During this process, more carbon dioxide is released and diffuses out of the mitochondrion and out of the cell. More hydrogens are released and picked up by NAD and another coenzyme called FAD. This produces reduced NAD and reduced FAD. At one pooint, ATP is produced.

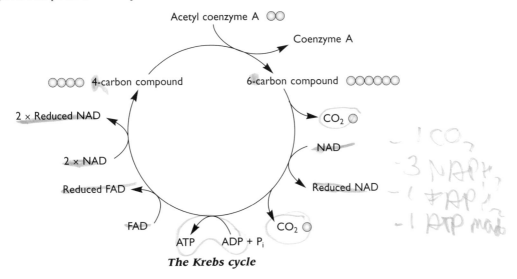

The Krebs cycle

Oxidative phosphorylation

The hydrogens picked up by NAD and FAD are now split into electrons and protons. The electrons are passed along the electron transport chain, on the inner membrane of the mitochondrion.

As they move along the chain, they lose energy. This energy is used to actively transport hydrogen ions from the matrix of the mitochondrion, across the inner membrane and into the space between the inner and outer membranes. This builds up a high concentration of hydrogen ions in this space.

The hydrogen ions are allowed to diffuse back into the matrix through special channel proteins that work as ATPases. The movement of the hydrogen ions through the ATPases provides enough energy to cause ADP and inorganic phosphate to combine to make ATP.

The active transport and subsequent diffusion of the hydrogen ions across the inner mitochondrial membrane is known as **chemiosmosis**.

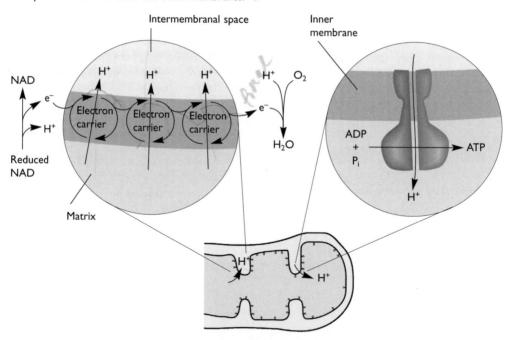

Oxidative phosphorylation

At the end of the chain, the electrons reunite with protons. They combine with oxygen, to produce water. This is why oxygen is required in aerobic respiration — it acts as the final acceptor for the hydrogens removed from the respiratory substrate during glycolysis, the link reaction and the Krebs cycle.

Anaerobic respiration

If oxygen is not available, oxidative phosphorylation cannot take place, as there is nothing to accept the electrons and protons at the end of the electron transport chain. This means that reduced NAD is not reoxidised, so the mitochondrion quickly runs out of NAD or FAD that can accept hydrogens from the Krebs cycle reactions. So the Krebs cycle and the link reaction come to a halt.

Glycolysis, however, can still continue, so long as the pyruvate produced at the end of it can be removed and the reduced NAD can be converted back to NAD. In animals, this is done by converting the pyruvate to lactate.

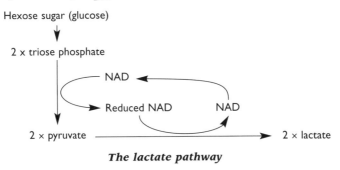

Hexose sugar (glucose)

2 × triose phosphate

NAD

Reduced NAD   NAD

2 × pyruvate ————————————————→ 2 × lactate

The lactate pathway

The lactate that is produced (usually in muscles) diffuses into the blood and is carried in solution in the blood plasma to the liver. Here, liver cells convert it back to pyruvate. This requires oxygen, so extra oxygen is required after exercise has finished. The extra oxygen is known as the oxygen debt. Later, when the exercise has finished and oxygen is available again, some of the pyruvate in the liver cells is oxidised through the link reaction, the Krebs cycle and the electron transport chain. Some of the pyruvate is reconverted to glucose in the liver cells, and this may be released into the blood or converted to glycogen and stored.

Activity of the heart

The heart is made of a specialised muscle tissue called cardiac muscle. This is **myogenic** — that is, it contracts and relaxes automatically, without the need for stimulation by nerves. The rhythmic contraction of the cardiac muscle in different parts of the heart is coordinated through electrical impulses passing through the cardiac muscle tissue.

- In the wall of the right atrium, there is a patch of muscle tissue called the sinoatrial node (**SAN**). This has an intrinsic rate of contraction a little higher than that of the rest of the heart muscle.
- As the cells in the SAN contract, they generate action potentials which sweep along the muscle in the wall of the right and left atria. This causes this muscle to contract. This is atrial systole (*Unit 1 Student Unit Guide* page 18).
- When the action potentials reach the atrioventricular node (**AVN**) in the septum, they are delayed briefly. They then sweep down the septum between the ventricles, along fibres called **bundles of His**, and then up through the ventricle walls. This causes the ventricles to contract slightly after the atria. The left and right ventricles contract together, from the bottom up. This is ventricular systole.
- There is then a short delay before the next wave of action potentials is generated in the SAN. During this time, the heart muscles relax. This is diastole.

Electrocardiograms

This electrical activity taking place in the heart can be monitored and recorded as an electrocardiogram or ECG.

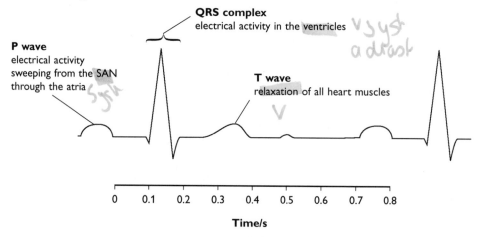

A normal ECG

Abnormal ECGs can be used to diagnose problems with the heart. For example:

- an unusually long delay between the P and R waves can mean that electrical impulses cannot pass easily from the atria to the ventricles, perhaps because of damage to the bundles of His
- a flat T wave can indicate lack of blood flow to cardiac muscle in the heart wall (ischaemia), so that it does not have energy for contraction. This could be an indication of cardiovascular disease (CVD).

Physiological effects of exercise

Exercise involves the contraction of muscles. Muscles obtain energy for contraction from ATP, which is produced by respiration. Exercise therefore requires an increased rate of respiration in muscle tissues, which in turn requires fast delivery of oxygen (and, to a lesser extent, glucose). This is achieved through:

- faster and deeper breathing, which increases the rate at which oxygen enters the blood in the lungs and carbon dioxide leaves it
- faster and stronger heart beat, which increases the rate at which blood moves through the blood vessels, delivering oxygen to muscle tissues and removing carbon dioxide and lactate from them

Control of heart rate

The **cardiac output** is the volume of blood leaving the left ventricle with each beat multiplied by the number of beats per minute. Cardiac output can be increased either by increasing the strength of contraction of the heart muscle (which increases the volume of blood pushed out with each beat) or by increasing the number of beats per minute. Usually, both occur.

Before $\rightarrow$ $\uparrow$ $\rightarrow$ CVC $\rightarrow$ Motor SAN $\rightarrow$ $\uparrow$ cah
Sympath

Just before and during exercise, the following events occur. NA

- **Adrenaline** is secreted from the adrenal glands. This stimulates the SAN to increase its rate of contraction.
- Action potentials are sent along a motor neurone (part of the **sympathetic** nervous system) from the **cardiovascular control centre** in the brain, to the SAN. The neurone releases a neurotransmitter called **noradrenaline** at the point where it reaches the SAN, causing the SAN to increase its rate of contraction.
  This may happen if the carbon dioxide concentration in the blood increases. This decreases the blood pH, because carbon dioxide forms a weak acid in solution. The low pH stimulates the cardiac centre to generate a higher frequency of action potentials in the sympathetic nerve. *chemo*
- When oxygen concentration falls in muscles (for example when they are using it up rapidly in respiration), the walls of the blood vessels in the muscles secrete **nitric oxide**. This makes the muscles in the arteriole walls relax, so the arterioles dilate and carry more blood. This increases the rate at which blood is returned to the right atrium of the heart, stretching the muscle in the atrial walls. The heart muscle responds to this by contracting more forcefully. The stretching also stimulates the SAN to contract faster. *Baro*

Control of ventilation rate

- During normal breathing, rhythmic patterns of nerve impulses are sent from the **ventilation centre** in the medulla oblongata in the brain to the muscles in the diaphragm and to the intercostal muscles, which respond by contracting rhythmically.
- Stretch receptors in the lungs are stimulated during breathing in, and send impulses to the ventilation centre, which uses this information to help to regulate breathing rate.
- During exercise, **chemoreceptors** in the ventilation centre sense a fall in pH, caused by increased carbon dioxide in the blood.
- Carbon dioxide concentration is also sensed by receptors in a patch of tissue in the wall of the aorta, called the **aortic body**.
- Another set of chemoreceptors in the walls of the carotid arteries, called **carotid bodies**, sense oxygen concentration, as well as carbon dioxide concentration, in the blood.
- Nerve impulses are sent from the aortic body and carotid bodies to the ventilation centre, and this then sends impulses to the breathing muscles, causing them to contract harder and faster. This increases the rate and depth of breathing.

Using a spirometer to investigate ventilation rate

A spirometer has an enclosed chamber containing air or medical-grade oxygen, lying over water. The lid of the spirometer can move up and down as the volume of air inside it increases and decreases. These movements can be recorded with a pen on a chart attached to a revolving drum.

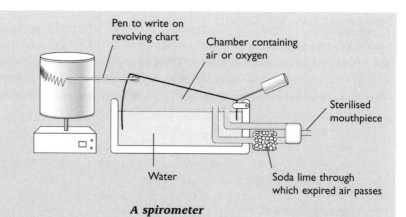

A spirometer

As a person breathes in and out through the mouthpiece, the lid of the air chamber goes down and up. This produces a trace on the chart.

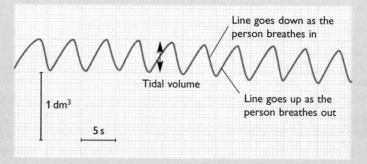

A spirometer trace

By counting the number of complete ups and downs in a known period of time, we can calculate the breathing rate in breaths per minute. In the example in the diagram, this is 12 breaths per minute.

The volume breathed in with each breath is the **tidal volume**. If we measure each of these, add them all and divide by the total number of breaths, we can find the mean tidal volume.

This allows us to calculate the ventilation rate:

ventilation rate = number of breaths per minute × mean volume of each breath

In the example in the diagram, mean tidal volume is $0.5\,dm^3$ and breaths per minute is 12, so the ventilation rate is $6\,dm^3\,min^{-1}$.

If spirometer traces are made before and after (or during) exercise, the effect of exercise on tidal volume and number of breaths per minute can be found.

Temperature regulation

Whenever energy is transferred from one place to another, or transformed from one type to another, some is always lost as heat. Respiration therefore produces heat, as does muscle contraction. During vigorous exercise, considerable quantities of heat are generated in muscles, and this causes the temperature of the blood to rise.

It is important that core body temperature remains roughly constant, around 37.8°C, as if it rises much above this then enzyme molecules may become denatured, so that metabolic reactions no longer take place normally and cells may be damaged. The regulation of core body temperature is part of **homeostasis** — the maintenance of a constant internal environment.

If core temperature rises

Any rise in temperature of the blood is detected by **thermoreceptors** in the **hypothalamus** in the brain. The hypothalamus also receives inputs from temperature receptors in the skin. These monitor the temperature of the external environment, and so can give 'early warning' that the body temperature may be going to rise or fall.

When a rise in temperature is detected, the hypothalamus sends nerve impulses to various effectors, which respond by taking action.

- **Arterioles** delivering blood to surface capillaries in the skin. Smooth muscle in their walls relaxes, causing them to dilate (vasodilation) so that a greater volume of blood flows into the surface capillaries. This allows heat to be lost by radiation from the blood through the skin surface.
- **Sweat glands** and the arterioles supplying them. More blood is delivered to the sweat glands, and they increase sweat production. Sweat flows up the sweat ducts onto the surface of the skin, where the water in it evaporates. This takes heat from the skin.
- **Erector muscles** attached to hairs. These relax, causing the hairs to lie flat, so they do not trap a layer of insulating air. This allows heat to radiate more easily from the skin to the surroundings.

If core temperature falls

- **Arterioles** delivering blood to surface capillaries in the skin constrict. This greatly reduces (and in extreme cold may completely stop) blood flow near the skin surface, diverting blood beneath the insulating fat (adipose tissue) layer beneath the skin, so that less heat is lost from it.
- **Sweat glands** secrete little or no sweat.
- **Erector muscles** contract, pulling hairs up on end. In furry animals (but not humans) this traps a layer of insulating air, greatly reducing heat loss from the skin.
- Certain **muscles** contract and relax very rapidly (shivering), generating extra heat which increases blood temperature.

Negative feedback

This mechanism of temperature regulation is an example of a negative feedback mechanism. In negative feedback, a receptor detects any change in the normal state of a system. If it detects an increase, it triggers events which bring about a decrease. It if detects a decrease, it triggers events which bring about an increase. The change is then detected by the receptor, which once again acts accordingly. The process is ongoing.

This type of system does not keep the state absolutely steady. In temperature regulation, for example, the core temperature does rise and fall a little. This is largely because of the time lag between the sensor (in this case the hypothalamus) detecting the change and the effectors (muscles in arterioles and so on) responding to it. So the temperature fluctuates a little, but rarely deviates very far from the norm. This is called a **dynamic equilibrium**.

Long-term temperature regulation

If a person spends a few days in a very cold environment, the hypothalamus releases greater quantities of a hormone called TRH, which stimulates the anterior pituitary gland to secrete TSH. This stimulates the thyroid gland to secrete more thyroxine.

Thyroxine travels in the blood to its target cells, where it diffuses through the cell surface membrane and into the nucleus. Here it switches on several genes which are responsible for encoding respiratory enzymes, especially cytochrome oxidase and ATPase. It also causes more mitochondria to be produced. This increases the rate of aerobic respiration in the cells, generating more heat.

DNA transcription factors

Thyroxine affects protein synthesis in a cell by binding to **transcription factors** in the nucleus of a cell. The activated transcription factors bind to specific regions of DNA (genes), and either increase or decrease the ability of RNA polymerase to attach to the DNA and catalyse the production of a complementary strand of mRNA from that gene. This may increase the transcription of a particular gene, called up-regulating, or it may decrease transcription, called down-regulating. Most steroid hormones, such as oestrogen and testosterone, act in this way.

Transcription factors may bind with a large number of different areas of DNA, so they can switch many different genes on or off. Thyroxine, for example, is known to affect the expression of at least 20 genes.

Exercise and health

Regular exercise increases fitness. For example:

- It decreases the risk of **obesity** (being seriously overweight), as it increases metabolic rate during the exercise — and there is often a temporary increase for some time after the exercise has finished. This helps to use fuels (respiratory substrates) such as glucose and fatty acids, reducing the amount of adipose tissue. Obesity is strongly linked to the development of type II diabetes, so keeping weight

down greatly lessens the risk of developing this disease.
- **Type II diabetes** is caused by a decrease in the sensitivity of liver and muscle cells to insulin. This means that high blood glucose levels are not returned to normal as fast as they should be. This can damage cells in all parts of the body.
- **Coronary heart disease** (CHD — see *Unit 1 Student Unit Guide*) is more likely to develop in people who do not exercise.

For all three of these conditions, studies have shown correlation between taking exercise and lowering the risk of developing the condition. There is also considerable reason to suspect a causal relationship, because understanding of the physiology involved suggests how this may be brought about.

However, people who exercise very frequently or who push their bodies to the limit also run the risk of causing damage to the body.
- Joints may become abnormally worn. For example, runners may damage knee joints by wearing away cartilage.
- There is evidence that very strenuous exercise taken over long periods of time can cause the immune system to become less effective. Athletes who put themselves under a lot of competitive stress and train very hard may be more likely to suffer infections than others. Laboratory studies repeatedly find a correlation between degree of exercise and the frequency of getting an infection of the upper respiratory tract (such as a cold). Moderate levels of exercise reduce the risk, but very high levels — 'overtraining' — decrease the activity of lymphocytes (especially T lymphocytes) and therefore the ability of the immune system to destroy viruses and other pathogens. The amount of an antibody called IgA, which is especially important in preventing infection of the upper respiratory tract, is also reduced by strenuous exercise, especially if this takes place over a long time period.

Treating injuries to joints
A relatively common injury in footballers is a torn anterior cruciate ligament. This ligament helps to support the bones at the knee joint.

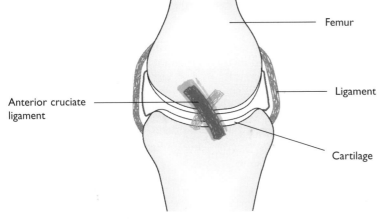

Femur

Ligament

Anterior cruciate ligament

Cartilage

The knee joint

This injury is generally treated using keyhole surgery, which greatly reduces the recovery time compared with normal surgery. One or more small holes are made leading into the knee joint, through which an arthroscope can be fed. (An arthroscope is an instrument that allows images of the inside of the joint to be observed by passing light rays along optical fibres to the observer's eye.) Surgical instruments are also passed through the holes. The surgeon will use tendons or ligaments taken from elsewhere in the body to replace the damaged ligament.

If a knee joint is very badly damaged, then the whole knee joint may be replaced with a prosthetic (artificial) joint. There are very many different types of these, but all are expensive as they must be made of high-quality materials that:
- are unaffected by body fluids
- can stand up to the high forces experienced by the knee
- have the right combination of flexibility and strength to allow free movement yet support the whole body weight

Use of performance-enhancing substances in sport

Professional athletes train hard to improve their performance. Some have been tempted to take a short-cut by taking performance-enhancing drugs. For example, steroids increase protein synthesis in cells and so can increase muscle size and strength. Erythropoetin (epo) increases the rate at which red blood cells are made, and so can increase the oxygen-carrying capacity of the blood.

Many performance-enhancing drugs are now banned — professional athletes face severe penalties imposed on them if they are found to have used them. This process is coordinated by the World Anti-Doping Agency. You can find up to date information about banned substances at **www.wada-ama.org/en**. The results of tests done on UK athletes can be found at **www.uksport.gov.uk/pages/drug_free_sport/**

There are very strong arguments for preventing the use of performance-enhancing substances in sport. These include:
- Many of these substances can cause damage to the athlete's health, especially if used over a long period of time. For example, they may cause liver damage. In some cases, they are thought to have led to the early death of an athlete, often by causing heart attacks.
- Athletes who use drugs may have a competitive advantage over those who do not, making competition unfair.

On the other hand, some people think that at least some types of performance-enhancing drugs should be allowed. Their arguments include:
- It is impossible to detect every performance-enhancing drug that can be used in sport. New drugs are always being tried, and it is difficult for regulators to develop new tests quickly enough to keep up with these developments.
- There is no ban on nutritional supplements such as vitamins — so where do we draw the line between an illegal performance-enhancing drug and a legal vitamin supplement?

Grey matter

Coordination

In multicellular organisms, such as plants and animals, it is essential that cells can communicate with each other. This allows them to coordinate their activities appropriately. Organisms have specialised cells or molecules that are sensitive to changes in their environment, called **receptors**. These trigger events in the organism that bring about coordinated responses to the environmental changes.

Plant photoreceptors

Light is of great importance to plants. They use it as their energy source in photosynthesis. Light also gives plants information about the seasons, to which they can respond by moving into different stages of their life cycles.

Plants respond to:
- light intensity
- light direction
- light quality (the particular mixture of wavelengths)
- light duration (daylength)

Plants have two main types of **photoreceptors** — that is, molecules that are changed when they are hit by light. These are phytochromes and blue-light receptors.

Phytochromes

Phytochromes are proteins that are sensitive to light with wavelengths between 600 nm and 800 nm — that is, red light. They are found in plant leaves. They exist in two forms, which are changed from one form to the other when they absorb light. Phytochrome red, P_r, absorbs red light (wavelength about 660 nm) and is changed to phytochrome far-red, P_{fr}, which absorbs far-red light (wavelength about 730 nm). This changes it back to P_r.

$$P_r \xrightleftharpoons[\text{Far-red light}]{\text{Red light}} P_{fr}$$

White light (including sunlight) contains both red and far-red light. In balance, this causes more P_{fr} to be produced than P_r. In the dark, P_{fr} slowly changes back to P_r. Therefore:
- After exposure to normal daylight, there is more P_{fr} than P_r.
- After a period in darkness, there is more P_r than P_{fr}.

In some species of plants, for example some types of lettuce, phytochromes affect seed germination. Dormant seeds of these plants will only begin to germinate when they contain plenty of P_{fr}, which happens when normal sunlight falls onto them. This is valuable because it prevents the seeds germinating in unsuitable conditions, i.e. where there would not be enough light for photosynthesis. If a light-sensitive seed is provided with the correct conditions for germination (i.e. water and a suitable temperature) and then is given a short burst of far-red light, it will not germinate. This is because the P_{fr} has been changed to P_r. If the burst of far-red light is followed by a burst of red light, however, the seed will germinate because the red light converts the P_r to P_{fr}.

Phytochromes also affect whether or not the plant produces flowers. Some plants are adapted to flower in spring, when the days are getting longer, and these are known as long-day plants. Others are adapted to flower in late summer or autumn, when the days are getting shorter, and these are known as short-day plants.

We have seen that, during darkness, P_{fr} changes to P_r. Short-day plants need a lot of P_r in their tissues in order to flower, which only happens when they have long, uninterrupted nights. If white or red light is shone on them even very briefly during the night, they will not flower, because this converts the P_r to P_{fr}.

Long-day plants need a lot of P_{fr} in their tissues in order to flower, which only happens when they have short nights — not long enough for all the P_{fr} to be converted to P_r.

Phytochromes exert their effects by activating other molecules in the plant cells which affect various metabolic pathways. Phytochromes also act as transcription factors in the nucleus, switching genes on and off.

Blue-light receptors

Plants also have receptors that respond to blue light, with a wavelength between 400 nm and 500 nm. These receptors were only discovered in the late 1980s and 1990s, and we still have a great deal to learn about them.

There are two types of blue-light receptors. **Cryptochromes** are involved in the change of a pale, non-photosynthesising seedling (just germinated and still mostly underground) to a green, photosynthesising one as it emerges into daylight. They sometimes act as transcription regulators, allowing or preventing the transcription of particular sets of genes. Cryptochromes also interact with phytochromes in the daylength-sensitive control of flowering.

Phototropins help to control the opening of stomata, which happens when bright light (containing blue wavelengths) falls onto a leaf. They also appear to be the main photoreceptors involved in phototropism — that is, the growth of a plant shoot towards the source of light.

Phototropism

Phototropism is a growth response by a plant to directional light. Shoots are **positively phototropic**, growing towards the source of light. Roots are often **negatively phototropic**, growing away from the light source.

This response involves a substance called **auxin** (of which **IAA** is one type). Auxin is a plant hormone, sometimes called a plant growth regulator. Auxin is made by cells in the growing point (meristem) of a plant shoot and then transported downwards through the shoot tissues. The presence of auxin in a cell activates transcription factors in the nucleus of the cell, switching different sets of genes on or off. Some of these genes code for the production of proteins called **expansins**. These act on cellulose cell walls, enabling them to stretch and expand when the cell takes in water. It is not yet certain how they do this, but it is probably by disrupting hydrogen bonds between cellulose molecules. Auxin therefore stimulates cell expansion.

The distribution of auxin in a shoot is affected by phototropin. When unidirectional light falls onto the shoot, this blue-light receptor affects the distribution of auxin, causing it to accumulate on the shady side. This causes the cells on the shady side to grow more rapidly than those on the well-lit side, so the shoot bends towards the light.

The nervous system of a mammal

Neurones
Neurones (nerve cells) are highly specialised cells that are adapted for the rapid transmission of electrical impulses, called **action potentials**, from one part of the body to another.

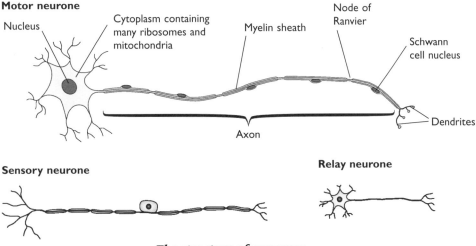

The structure of neurones

Information picked up by a receptor is transmitted to the central nervous system (brain or spinal cord) as action potentials travelling along a sensory neurone. These neurones have their cell bodies in small swellings, called ganglia, just outside the spinal cord. The impulse is then transmitted to a relay neurone, which lies entirely within the brain or spinal cord. The impulse is then transmitted to many other neurones, one of which may be a motor neurone. This has its cell body within the central nervous system, and a long axon that carries the impulse all the way to an effector (a muscle or gland).

In some cases, the impulse is sent on to an effector before it reaches the 'conscious' areas of the brain. The response is therefore automatic, and does not involve any decision-making. This type of response is called a **reflex**, and the arrangement of neurones is called a **reflex arc**.

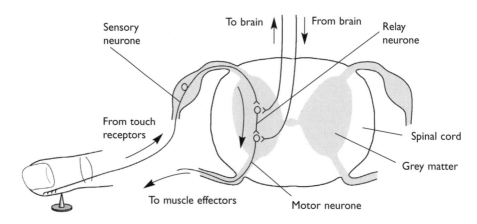

Arrangement of neurones in a reflex arc

The myelin sheath is made up of many layers of cell membrane of Schwann cells, which wrap themselves round and round the axon. This provides electrical insulation around the axon which, as we shall see, greatly speeds up the transmission of action potentials. Not all neurones are myelinated.

Action potentials

Neurones, like all cells, have sodium-potassium pumps in their cell surface membranes. However, in neurones these are especially active. They pump out sodium ions and bring in potassium ions, by active transport. Three sodium ions are moved out of the cell for every two potassium ions that are moved in.

There are also other channels in the membrane that allow the passage of sodium and potassium ions. When a neurone is resting, quite a few potassium ion channels are open, so potassium ions are able to diffuse back out of the cell, down their concentration gradient.

As a result, the neurone has more positive ions outside it than inside it. This means there is a potential difference (a voltage) across the axon membrane. It has a charge of about −70 mV (millivolts) inside compared to outside. This is called the **resting potential**.

When a receptor receives a stimulus, this can cause a different set of sodium channels to open. This allows sodium ions to flood into the cell, down an electrochemical gradient. (The 'electro' gradient refers to the difference in charge across the membrane. The 'chemical' gradient is the difference in concentration of sodium ions.) This quickly reverses the potential difference across the cell membrane, making it much less negative inside. The neurone is said to be **depolarised**. Indeed, the sodium ions keep on flooding in until the cell has actually become positive inside, reaching a potential of about +30 mV. The sodium ion channels then close.

This change in potential difference across the membrane causes a set of potassium ion channels to open. These are called voltage-gated channels, because they open when the potential difference (voltage) across the membrane is positive inside. Potassium ions can now flood out of the axon, down their electrochemical gradient. This makes the charge inside the axon less positive. It quickly drops back down to a little below the value of the resting potential.

The voltage-gated potassium ion channels then close. The sodium-potassium pump swings back into operation, and the resting potential is restored.

This sequence of events is called an **action potential**. The time taken for the axon to restore its resting potential after an action potential is called the **refractory period**. The axon is unable to generate another action potential until the refractory period is over.

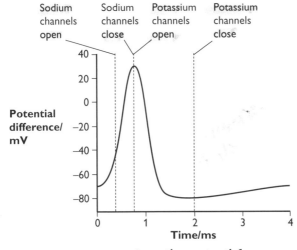

An action potential

Transmission of action potentials

An action potential that is generated in one part of a neurone travels rapidly along its axon or dendron. This happens because the depolarisation of one part of the membrane sets up local circuits with the areas on either side of it. These cause depolarisation of these regions as well. The action potential therefore sweeps along the axon.

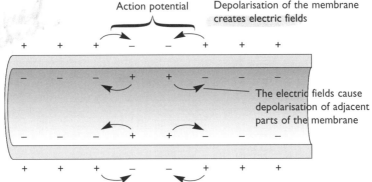

How an action potential travels along a neurone

In a myelinated neurone, local circuits cannot be set up in the parts of the neurone where the myelin sheath is present. Instead, the action potential 'jumps' from one node of Ranvier to the next. This greatly increases the speed at which it travels along the axon.

Synapses

saltatory junction

Where two neurones meet, they do not actually touch. There is a small gap between them called a **synaptic cleft**. The membrane of the neurone just before the synapse is called the **presynaptic membrane**, and the one on the other side is the **postsynaptic membrane**. The whole structure is called a **synapse**.

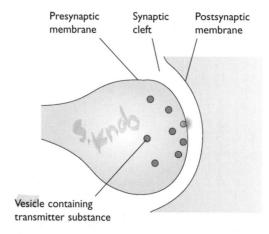

A synapse

- When an action potential arrives at the presynaptic membrane, it causes voltage-gated calcium ion channels to open.
- Calcium ions rapidly diffuse into the cytoplasm of the neurone, down their concentration gradient. *endo*
- The calcium ions affect tiny vesicles inside the neurone, which contain a **transmitter substance** such as **acetylcholine**. These vesicles move towards the presynaptic membrane and fuse with it, releasing their contents into the cleft.
- The transmitter substance diffuses across the cleft and slots into receptor molecules in the postsynaptic membrane.
- This causes sodium ion channels to open, so sodium ions flood into the cytoplasm of the neurone, depolarising it.
- This depolarisation sets up an action potential in the postsynaptic neurone.

Receptors in the eye

The retina of the human eye contains highly specialised cells that are receptive to light. There are two types — **rods**, which are sensitive to dim light but do not differentiate between different wavelengths, and **cones**, which only respond in bright light but which do respond differently to different wavelengths (colours) of light.

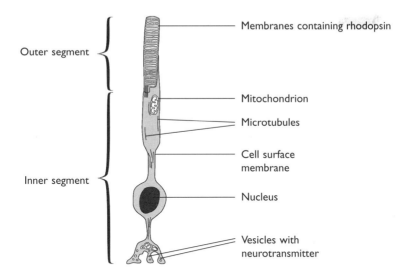

A rod cell

The outer segment of a rod cell contains many **membranes**, all stacked up in parallel to one another. Sodium-potassium **pumps act** across the membranes, as in most cells. As always, other potassium and sodium channels (cation channels) remain open, so some of the sodium and potassium ions leak back through to where they came from.

These membranes also contain a pigment called **rhodopsin**, whose molecules are made up of a retinal molecule and an opsin molecule. When light hits a rhodopsin molecule, the retinal part changes shape. This causes the sodium and potassium ion channels in nearby areas of the membrane to close, but the sodium-potassium pump keeps on working. As the sodium and potassium ions cannot now leak back in or out of the neurone, a greater potential than usual builds up across it (negative inside) because of the activity of the sodium-potassium pump. The membrane is said to be **hyperpolarised**.

When no light is falling on a rod cell, and the potential across its membrane is normal, it constantly releases transmitter substance, which diffuses to a neurone next to it. This transmitter substance stops that neurone generating action potentials. When light falls on the rod cell and its membranes become hyperpolarised, this causes it to stop releasing transmitter substance. Now the neighbouring neurone can generate action potentials. These are transmitted along axons in the optic nerve, which carry them to the visual centre in the brain.

The change in shape of retinal causes it to become unstable. After a minute or so, it separates from the opsin. If this happens to all the rhodopsin in all the rod cells, you no longer have any working rhodopsin and cannot see in dim light. This explains why, if you walk from a sunny street into a dimly-lit room, you cannot see much. In the dim light, the retinal and opsin gradually combine again, forming rhodopsin, so you become able to see once more. This is called **dark adaptation**.

The iris reflex

The **iris** is the coloured part of the eye, surrounding a gap called the **pupil**. Light passes through the pupil on its way to the retina. In bright light, the pupil is small (contracted), to limit the amount of light passing through, as very bright light can damage rods and cones. In dim light, the pupil is large (dilated), to allow more light to reach the retina.

When bright light hits the retina, receptors generate action potentials, which travel to the brain. This causes action potentials to be sent along motor neurones to the muscles in the iris. Circular muscles contract and radial muscles relax, making the iris wider and therefore the pupil narrower.

In dim light, the opposite takes place. Circular muscles relax and radial muscles contract, widening the pupil.

Comparison of coordination in plants and animals

- Coordination in both plants and animals involves receptors, a communication system and effectors.
- Animals have a nervous system, containing specialised neurones which transmit action potentials very rapidly. Plants do not have a nervous system or neurones. However, some parts of plants do transmit action potentials, but this is generally done very much more slowly than in animals, and the potential differences involved are generally less than those in mammals. For example, a fly touching the hairs on the leaf of a Venus's fly trap causes an action potential to sweep across the leaf, bringing about its closure.
- Both animals and plants also use chemicals that are produced in one part of the organism and travel to other parts where they have their effects. In animals, these substances are called hormones, and they are made in endocrine glands, which secrete them directly into the blood. They are then carried in solution in the blood plasma. They affect target organs which have receptors for them.
  In plants, however, there are no glands where these chemicals are made. Like animal hormones, plant hormones are made in one area (for example auxin is made in meristems) and travel to another part of the plant where they have their effect. However, unlike animal hormones, they often do not travel in vessels but instead move through cells, either by facilitated diffusion through protein channels or by active transport through protein transporters.
  Animal hormones are almost all small protein molecules or steroids. So far, no protein or steroid plant hormones have been found.

The human brain

The brain is a part of the central nervous system. Its role is to initiate and coordinate activities in different parts of the body. It receives input from receptors both inside and at the surface of the body, and the information from these is integrated to produce appropriate actions in response.

The brain is made up of neurones, and also other cells called glial cells. Different parts of the brain have different functions.

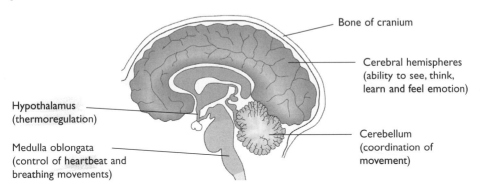

Bone of cranium

Cerebral hemispheres (ability to see, think, learn and feel emotion)

Hypothalamus (thermoregulation)

Cerebellum (coordination of movement)

Medulla oblongata (control of heartbeat and breathing movements)

The human brain

One way of investigating the functions of different parts of the brain is to use **functional magnetic resonance imaging, fMRI**. The person lies inside a huge cylindrical magnet. The magnetic field affects the nuclei of atoms inside the person's body, causing them to line up in the direction of the field. This sends a tiny signal to a computer, which builds up an image by mapping the strengths of the signals from different parts of the brain. Haemoglobin and oxyhaemoglobin produce different signals, so the scanner is able to map where there is most oxyhaemoglobin, which indicates where neurones are most active and therefore where most blood is flowing to. By comparing the images obtained when a person's brain is doing different things (such as reading a book, thinking about food and so on) the areas of the brain involved in these tasks can be worked out.

Another way of mapping brain activity is to use **computer-assisted tomography**, known as a **CAT scan**. This uses X-rays, which are sent through the body at different angles. Just like an ordinary X-ray, they can produce an image showing dense and less dense areas. A computer builds up a 3-dimensional image. CAT scans are much cheaper than fMRI scans, but they do not show which tissues are active, so they can only show structure and not function of the brain. They are very useful in the diagnosis of conditions such as brain tumours or encephalopathies, where changes in brain structure can be visualised.

Development of visual capacity
We do not 'see' a simple picture of the image that is formed on our retina. The brain processes this image, using past experience and other sensory inputs to enhance the information obtained. For example, if a tiny image of a person falls onto your retina, your brain does not 'see' this as an extremely small person standing just in front of you, but as a normal-sized person standing some distance away.

The capacity of the brain to process and interpret the action potentials that arrive along the optic nerve is acquired during early childhood. The brain learns through experience. Newborn babies have very little ability to interpret the information their eyes collect, but this is highly developed by the age of 5 or 6 years. The experiences that the child has help to determine the way the brain cells are 'wired up' during this development.

In the 1960s, two researchers in the USA, Hubel and Wiesel, carried out some important experiments using monkeys and kittens. They found that if they prevented light reaching the retina of one or both eyes as the young animals matured, this stopped them developing normal visual abilities. Depriving them of vision at different stages of development affected different aspects of vision. They concluded that there were specific 'windows' during development, which they called **critical periods**, during which particular types of visual input were needed in order for particular visual capacities to develop normally.

Since then, much more information has been obtained, including some from human babies. Babies are sometimes born with cataracts, which prevent light patterns from reaching the retina. Sometimes, the baby is born with normal eyes but a cataract develops during the first few years of life. These cataracts can be removed so that vision becomes normal. These babies are therefore deprived of normal visual input at different periods of their development. The researchers have tested their visual abilities as they grow up, and looked for correlations between the time of visual deprivation and any particular visual abilities that fail to develop properly.

They have found that there is not just one 'critical period' for development of a particular visual ability. Instead, there are three different types of period:
- a sensitive period when there are developmental changes in the eyes and brain that do not occur if a particular visual input is not experienced
- a period after this first sensitive period when, even if the visual ability has developed normally and fully, it can be damaged if abnormal visual input is experienced
- a period after this when any damage caused by earlier deprivation can be reversed if normal visual input is given

For example, one visual ability is called visual acuity. This is the ability to distinguish objects of small sizes. It is usually tested by a person's ability to tell the difference between a plain grey square and a square divided up by stripes. The narrower the stripes that can be distinguished, the better the visual acuity.

Researchers have found that, in human babies:
- a newborn baby's visual acuity is very poor, around 40 times worse than in adults
- visual acuity improves rapidly during the first 6 months if both eyes are working normally during this time period, and reaches adult values at around 5 years of age
- babies born with cataracts that are not treated until 9 months old (i.e. have no visual input until then) have the visual acuity of a newborn baby when their cataracts are removed

- these babies then improve their visual acuity rapidly, so that by 1 year old they have the same visual acuity as a baby that has developed normally
- they then fall behind in their development, so that by 5 years old their visual acuity is 3 times worse than normal, and they never do develop normal adult visual acuity

This indicates that there is no one 'critical window' when visual acuity develops, but rather a range of 'windows' in which deprivation of normal visual input has different effects on development.

Similar results, but covering different time periods, have been obtained for other visual abilities, such as binocular vision and peripheral vision.

The 'nature versus nurture' debate

There has long been an interest in the extent to which the behaviour of animals (including humans) is determined by genes or by the environment — that is, the experiences an animal has as it develops. The effect of genes is sometimes known as 'nature' and the effect of the environment as 'nurture'. Today, we understand that most patterns of behaviour are determined by both genes and environment.

A behaviour pattern that is shown at the very beginning of an organism's life is said to be **innate**, and is considered to be caused entirely by genes. However, in humans, even a newborn baby has already been influenced by its environment, as it was developing in the uterus. Newborn babies show various reflex actions, such as the 'startle reflex'. This happens when the baby hears a sudden loud noise, or is dropped a short distance. The baby responds by flinging out the arms and legs and contracting the neck muscles. It is likely that this response is largely innate (caused entirely by genes) but it may also have been influenced by the experiences of the baby while it was a fetus in the uterus.

Brain development and behaviour are influenced by both genes and the environment, with no sharp dividing line between innate behaviour and learned behaviour. One way of investigating this is to study **identical twins**. Identical twins have identical genes. The brain development and behaviour of identical twins who have been brought up in different families and environments can be compared. Any differences between them must have been influenced by their different environments.

Animal studies confirm that innate patterns of behaviour can be modified by experience. For example, the females of an ichneumon fly lay their eggs in caterpillars, and the larvae then feed on the caterpillar from the inside. The females always lay their eggs in the same species of caterpillar. This appears to be innate behaviour, programmed by genes. However, if some of the larvae are placed in a different species of caterpillar, then when they grow into female ichneumon flies these lay their eggs inside the same kind of caterpillar in which they developed. The basic egg-laying behaviour, which is innate, has been modified by the ichneumon flies' environment.

Habituation

Even the simplest organisms modify their behaviour as a result of experience. This is called learning. One of the simplest types of learning is **habituation**. This can be defined as a decrease in the intensity of a response when the same stimulus is given repeatedly.

For example, a sea slug called *Aplysia* withdraws its gills when it is touched. This response helps it to avoid damage by predators. If it is touched repeatedly, and nothing unpleasant happens to it, it eventually stops withdrawing the gills. This is useful as it avoids energy being wasted on an unnecessary action, and enables the sea slug to stay fully active.

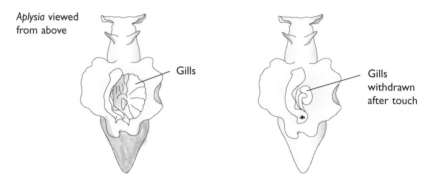

Aplysia viewed from above

Gills

Gills withdrawn after touch

Sea slug Aplysia

Humans also show habituation. For example, if you hear a loud bang outside the room, you may initially respond by jumping and then going to find out what it is. If the bangs keep happening all day, then eventually you will not only stop responding to them, but you may even not notice them after a while.

Investigating habituation in pond snails

Collect several pond snails of the same species. Place them in a tank containing pond water and weeds, well aerated. Leave them for a few days to acclimatise.

Remove one snail from the tank and place it in a dish of water taken from the tank. Leave the snail for at least five minutes. The snail must be active when you do your experiment.

Using a seeker or other small implement, gently touch the snail on the tip of one tentacle. (The tentacles have eyes on the end, so it is important not to damage these.) The snail should withdraw its tentacle and then slowly extend it again. Repeat the stimulus several times, with set time intervals of less than one minute between each stimulus. Record the time taken for the tentacle to be returned to its fully extended position after each stimulus. Plot a graph of time taken for the tentacle to be withdrawn and re-extended against number of stimuli given.

You can extend your experiment to investigate other aspects of habituation. For example, if you test the same snail again on a subsequent day, is the pattern of response the same? If you test the other tentacle, does the snail transfer what it has learned from one tentacle to the other?

Issues relating to the use of animals in research

Arguments against using animals in research include:

- We have no right to submit animals to procedures that may cause them discomfort or make their lives unpleasant.
- There is no need to use animals in research, as there are other ways of conducting the same investigations.

The main argument for using animals in research is:

- This may be the only way we can fully test new drugs and other substances, or find out more about a particular aspect of physiology or behaviour, which may eventually lead to less suffering for humans and other animals.

All institutions in the UK that use animals in research follow codes of conduct which:

- limit the use of animals in research to circumstances in which no alternative method (such as using cells grown in tissue culture or using human volunteers) is possible
- only allow research using animals to be carried out after thorough scrutiny of the researcher's proposal, which must show that no alternative is possible and that the welfare of the animals will be given high priority at all times
- stipulate that all people involved in the use of animals in research, including scientists and those who care for the animals, are given full training in ensuring the health and wellbeing of the animals

Drugs and the brain

Neurotransmitters in the brain

The brain contains a huge number of neurones, which transmit nerve impulses to each other across synapses. Many different neurotransmitters are involved, including dopamine and serotonin.

Dopamine is produced in several parts of the brain. One of these areas is the substantia nigra. Impulses passing between neurones in the substantia nigra are important for controlling movement. Parkinson's disease occurs when cells in the substantia nigra die, so dopamine is no longer produced, resulting in loss of control of movement.

Parkinson's disease is treated using a drug called levodopa or **L-Dopa**. This enters the brain, where the enzyme dopa decarboxylase converts it to dopamine. (It is not possible to give dopamine itself as a drug, as it cannot enter the brain.) This treatment is very helpful to people in the early stages of Parkinson's disease, but the dose has to be increased as the disease progresses, because the cells that convert L-Dopa to dopamine gradually die. It is also important not to give too much L-Dopa, as this can give rise to uncontrolled movement.

Serotonin (also known as 5-hydroxytryptamine or 5-HT) occurs as a neurotransmitter in many parts of the brain. It appears to be involved in many different functions, including mood, appetite, temperature regulation, sensitivity to pain and sleep. Lack of serotonin has been linked to the development of clinical depression, although most researchers consider that clinical depression has a more complex underlying cause than a simple lack of serotonin. Clinical depression can be treated with drugs that increase the quantity of serotonin in the brain.

The drug **MDMA**, generally known as **Ecstasy**, also acts at serotonin synapses in the brain. It binds to a transporter protein which normally removes serotonin from the synapse. This means that the effects of serotonin at synapses are increased. MDMA can produce feelings of euphoria, friendliness and energy. However, it has many other effects in the body, and it can also cause depression, confusion and anxiety. Animal research shows that regular use of MDMA causes damage to several parts of the brain.

The Human Genome Project and drug development

The Human Genome Project has worked out the base sequences of all the DNA in a human cell. This has led to a knowledge of the base sequences of tens of thousands of genes, many of which we did not previously know existed. From this, we can work out the amino acid sequences of the proteins that they produce, and this can lead to an understanding of how these proteins work. This in turn may enable researchers to develop new drugs that specifically target particular proteins, either enhancing their activity or lessening it.

We are also beginning to obtain information about different alleles of particular genes. When drugs are being used to treat disease, not everyone responds to the same drug in the same way, and knowledge of differences in the person's base sequences in particular genes can help us to understand this. It is envisaged that, in the future, knowledge of a person's particular DNA sequences will enable suitable drugs to be chosen on an individual basis.

Using genetically modified organisms to produce drugs

Genes for the synthesis of particular proteins can be inserted into the DNA of an organism. This genetically modified organism (GMO) may then express that gene and synthesise the protein.

The production of a genetically modified organism involves:

- identifying and isolating the gene that is to be inserted into it; this can be done by cutting the gene from DNA using restriction enzymes, or by 'reverse engineering' by using the sequence of amino acids in the protein to be made and working out and constructing a length of DNA with the appropriate base sequence to code for this protein
- inserting the gene into a vector, such as a bacterial plasmid or a virus
- inserting the vector into the organism which is to be modified

To ensure that the inserted gene will be expressed, it may also be necessary to insert a length of DNA called a promoter, which is required in order for RNA polymerase to begin transcription of the gene.

Examples of genetically modified organisms currently being used for drug production include:

- tobacco plants that produce an anti-inflammatory cytokine called interleukin-10. This may be used to treat autoimmune diseases.
- *Escherichia coli* bacteria that produce insulin, used for the treatment of diabetes
- maize plants that produce human lipase, which may be used to treat people with cystic fibrosis whose pancreatic juice (containing lipase) is insufficient
- cultured cells taken from hamsters, which produce erythropoetin (a hormone which increases the production of red blood cells)
- goats that produce antithrombin (a drug sold as ATryn®) in their milk, which is used to treat people with a blood clotting disorder

Benefits and risks of GMOs

Potential benefits of using genetically modified organisms include:

- Crop plants can be produced that are resistant to attack by certain pests, reducing the need to use pesticides. This can increase yields and reduce the risk of harming beneficial insects.
- Crop plants can be produced that are resistant to a particular herbicide, allowing the herbicide to be sprayed on the crop where it will kill weeds but not the crop plants.
- Crop plants can be modified so that they produce higher quantities of a particular nutrient, for example Golden Rice, which produces β-carotene (a precursor of vitamin A) and therefore reduces the risk of children suffering from vitamin A deficiency.
- As we have seen, drugs can be produced in ways that do not involve harvesting them from an organism (for example, in the past insulin was obtained from the pancreases of pigs) or by chemical synthesis (which can be very expensive).

Potential risks include:

- Genes inserted into a crop plant might spread to others in pollen. This could cause changes in the genotypes of wild plant populations, which could adversely affect other organisms in the ecosystem. For example, genes from crops engineered to be resistant to insects could spread into wild plants, so that insects would no longer have a good food supply. There has so far been limited evidence that this has ever occurred, but it is a genuine risk.
- Pests might develop resistance, through natural selection, to the substance in GM crops that confers resistance to pests; this could result in a population of 'super-pests'. This is a genuine risk, but it is probably no greater than the risk resulting from pests developing resistance to pesticides.
- Some people consider that consuming foods made from or containing GMOs could be harmful to health. There is no evidence so far that this is a genuine risk.

Questions
&
Answers

In this section there are two sample examination papers, similar to the Edexcel Unit Test papers. All of the questions are based on the topic areas described in the previous sections of the book.

You have 1 hour 30 minutes to do each paper. There are 90 marks on the paper, so you can spend almost 1 minute per mark. If you find you are spending too long on one question, move on to another that you can answer more quickly. If you have time at the end, come back to the difficult one.

Some of the questions require you to recall information that you have learned. Be guided by the number of marks awarded to suggest how much detail you should give in your answer. The more marks there are, the more information you need to give.

Some of the questions require you to use your knowledge and understanding in new situations. Don't be surprised to find something completely new in a question — something you have not seen before. Just think carefully about it, and find something that you *do* know that will help you to answer it.

The final question in each paper is based on a scientific article. You will receive this article some days before the examination. The scientific articles for these sample papers are on pages 47–54 and 55–60. They are shorter than the one you will receive for the 'real' paper, but they will give you a good idea of how to answer this style of question.

Do think carefully before you begin to write. The best answers are short and relevant — if you target your answer well, you can get many marks for a small amount of writing. Don't ramble on and say the same thing several times over, or wander off into answers that have nothing to do with the question. As a general rule, there will be twice as many answer lines as marks. So you should try to answer a 3-mark question in no more than 6 lines of writing. If you are writing much more than that, you almost certainly haven't focused your answer tightly enough.

Look carefully at exactly what each question wants you to do. For example, if it asks you to 'Explain', then you need to say *how* or *why* something happens, not just *what* happens. Many students lose large numbers of marks by not reading the question carefully.

Following each question, there is an answer that might get a C or D grade, followed by an examiner's comments, indicated by *e*. Then there is an answer that might get an A or B grade, again followed by examiner's comments. You might like to try answering the questions yourself first, before looking at these.

Sample paper

Scientific article 1
Bridging the gap

The first experiments on the role of electric currents in the transmission of nerve impulses were performed over 200 years ago by Luigi Galvani. Galvani showed that electrically stimulating nerves in dissected frog legs caused muscles to contract. Since then, our understanding of how impulses are transmitted along neurones has grown immensely. We now know that impulses happen as all-or-nothing waves of excitation, termed action potentials. Action potentials occur through the coordinated opening and closing of voltage-gated ion channels in the neuronal membrane of nerve cells. Ion channels are protein pores that allow ions such as Na^+, Ca^{2+}, K^+ or Cl^- to cross the membrane. The term 'voltage-gated' refers to the fact that these ion channels open when the electrical charge across the membrane changes.

However, in most situations where a neurone meets or forms a synapse with its target cell, there is no direct connection between the neurone and its target. Common sense tells us that without a direct connection, impulses cannot be conducted. This leaves us with a problem: how are impulses transmitted from neurone to target?

Chemical transmission

The riddle of how neurones signal to their targets was solved in 1921 by Otto Loewi. Loewi observed that if he stimulated a nerve called the vagus, a frog's heart rate would slow down. What Loewi then did was to collect the fluid surrounding the stimulated heart and apply it to a second, unstimulated heart. He was astonished to see that the second heart slowed, indicating that a chemical released by the vagus nerve had accumulated in the fluid he had applied. This substance was later shown to be acetylcholine. Loewi's experiments showed for the first time that neurones transmit signals to their targets by releasing substances that then diffuse across the synapse. These substances are called neurotransmitters and they are recognised by specific receptor proteins on the membrane of the target cell.

Although acetylcholine was the first neurotransmitter isolated, it is by no means the only such substance found in the nervous system. Other common neurotransmitters include gamma aminobutyric acid (GABA), glutamate, dopamine, serotonin and noradrenaline. As there are so many different chemical signals, it is important that receptors recognise only one kind of neurotransmitter. Selectivity is achieved by making the binding site for the neurotransmitter complementary to the structure of the neurotransmitter. This is similar to the way that an enzyme recognises its substrate. Selectivity can be exploited by pharmacologists because if they synthesise a drug, the structure of which is close to that of a naturally occurring neurotransmitter, it may well bind specifically to the receptor for that neurotransmitter.

**scientific
article**

Ligand-gated ion channels

Once a receptor protein has recognised its neurotransmitter, the information it carries has to be conveyed inside the target cell. There are several ways in which this can happen but one of the most common and important mechanisms is when the receptor has an ion channel built into its structure. This type of receptor is usually termed a ligand-gated ion channel. The term ligand means 'a molecule capable of binding', so a 'ligand-gated ion channel' is an ion channel the opening of which is controlled by the binding of a ligand, i.e. the neurotransmitter that the receptor recognises. When the neurotransmitter binds, it causes a change in the structure of the receptor that opens the channel and allows ions to flow into or out of the cell. This ion flow can change the behaviour of the target cell by altering the charge across the external surface of the cell, the plasma membrane.

Just like voltage-gated ion channels, ligand-gated channels are highly ion-selective and will only allow a limited range of ions to pass through. Some are selective for sodium ions and some are highly permeable to calcium ions. Other types only allow chloride ions (which have a negative charge) to enter. As we will see later, increased permeability to each of these three ions will affect the behaviour of the target cell in a different way.

The neuromuscular junction

The neuromuscular junction is the synapse between a motor neurone and skeletal muscle. It is one of the best understood and most studied examples of chemical transmission involving a ligand-gated ion channel. An action potential arriving at the terminal of the motor neurone triggers the release of acetylcholine into the synaptic cleft — the space separating the two cells. The acetylcholine then diffuses across to the muscle membrane where it binds to a receptor called the nicotinic acetylcholine receptor (often shortened to 'nicotinic receptor'). This ligand-gated ion channel, so-named because it is activated by nicotine, then opens and allows sodium ions to enter the muscle. This depolarises the muscle and triggers an action potential, which ultimately leads to muscle contraction. Finally, the signal is terminated by the action of acetylcholinesterase. This is an enzyme, found in the synaptic cleft, which breaks down acetylcholine into acetic acid and choline.

Nicotinic acetylcholine receptors

Electric rays are fish capable of delivering shocks of around 200 V that can kill or stun their prey. These discharges are produced by electric organs, which contain a specialised form of muscle. Electric organs contain large numbers of nicotinic receptors and so have provided scientists with a rich source of material with which to study the structure and function of the receptor. Biochemical studies of electric organ nicotinic receptors revealed that the receptor protein is made up of five smaller proteins (subunits). The subunits are arranged in a ring and form a tube through the membrane. It is this tube that forms the ion channel. Four types of subunit are found:

alpha, beta, gamma and delta. Each receptor contains two alpha subunits, one beta, one gamma and one delta. The five subunits each span the membrane with about 50% of their mass outside the cell. It is in this extracellular part of the receptor that the two binding sites for acetylcholine are found, sandwiched between the alpha subunits and their neighbouring gamma or delta subunits. In order for the channel to open, both binding sites must be occupied by acetylcholine. The binding of acetylcholine causes the shape of the protein to change so that the diameter of the channel increases and ions can pass through.

Drugs and toxins targeting nicotinic acetylcholine receptors

During surgery patients are often deliberately paralysed by surgical staff. This prevents unwanted movements and allows breathing rate to be controlled by artificial ventilation. Surgical paralysis is induced using two types of drugs that target muscle nicotinic receptors to prevent muscle contraction. The first type of drug is known as a depolarising blocker. Depolarising blockers have a structure similar to that of acetylcholine and can activate the nicotinic receptor. However, they are not broken down by acetylcholinesterase — the enzyme that normally terminates signalling at the neuromuscular junction. As a result, the receptor stays activated for longer than normal. This causes the muscle to become electrically inactive and also causes the receptor to enter a 'desensitised' state in which the ion channel closes. The second type of drug is known as a non-depolarising blocker. Such drugs bind tightly to the acetylcholine binding sites but do not activate the receptors. This means that acetylcholine itself cannot gain access to the receptor and so cannot trigger muscle contraction. Paralysis by depolarising and non-depolarising blockers is normally terminated by the body metabolising the drug (either in the liver or via enzymes in the circulation) or excreting it via the kidney, but some blockers break down spontaneously.

Nicotinic receptors are also affected by animal and plant toxins. Perhaps the most famous example is the arrow poison curare, which is obtained from plants of the *Strychnos* genus. Possession of such toxins has probably evolved to discourage animals from eating the plant. Curare acts as a non-depolarising blocker and so will paralyse animals shot with arrows that have been coated in it. It is, fortunately, not absorbed from the gut so such animals are safe to eat. A second example is alpha cobra toxin, found in cobra venom. Alpha cobra toxin is a peptide that binds almost irreversibly to the nicotinic receptor acetylcholine sites. Again, this produces non-depolarising blocker and so will immobilise the cobra's prey and ultimately kill it due to paralysis of the muscles involved in breathing.

Ligand-gated ion channels in the brain

Ligand-gated ion channels are also involved in chemical transmission in the brain but the receptors found there are different from those in muscle. At most synapses in the brain, action potentials are relayed from neurone to neurone by the

scientific article

neurotransmitter glutamate. Glutamate binds to several different types of ligand-gated ion channel called ionotropic glutamate receptors. Some of these receptors are highly permeable to calcium and are thought to be important in learning and memory. However, when part of the brain is deprived of oxygen, such as during a stroke, too much glutamate can sometimes be released and these receptors can become over-activated. This causes too much calcium to enter the neurones. High levels of calcium can trigger the death of neurones. This may lead to permanent brain damage.

A second type of ligand-gated ion channel important in the brain is the GABA-A receptor. This receptor binds to the neurotransmitter GABA and contains a chloride ion selective channel. Instead of triggering action potentials, making neurones more permeable to chloride renders them less excitable and so GABA is an inhibitory neuro-transmitter. Many compounds have been discovered that make GABA-A receptors work better, thereby decreasing activity in the brain. Drugs that work by this mecha-nism include sleeping tablets such as temazepam, anti-anxiety drugs such as valium and certain general anaesthetics.

Ligand-gated ion channels are the body's molecular switches. They are involved in the operation of every level of the nervous system. Investigating their structure and function will not only give us a better understanding of how the nervous system works, but also allow the development of better drugs to treat a wide range of disorders.

Source: Biological Sciences Review, *April 2008*

Biological clocks

Most organisms experience marked day–night cycles, but above the Arctic Circle day and night occur only in spring and autumn. In summer there is sunshine around the clock and in winter there is only a short period of twilight at noon. Further north, at the high-Arctic islands of Svalbard (Spitzbergen is one of them), there is permanent darkness ('polar night') from November till February. Permanent light ('polar day') lasts from April till September. Some of these islands now have people living on them. Terrestrial animals inhabited the islands when the ice retreated during the last ice age — reindeer arrived about 5000 years ago. All the animals and plants arrived from much further south, carrying in their genes a biological clock set to function at lower latitudes.

Biological clocks and rhythms

Most animals show rhythmic daily patterns. People, cows and many birds rest during the night. Most rodents and owls are active at night and rest during the day. Plants, too, show daily rhythms, such as flowers opening and the capacity to photosynthesise (higher during daylight hours). Many functions change in parallel with this cyclic behaviour. For example, our body temperature is lower at night, and growth hormone secretion increases when we sleep. Such biological rhythms represent a fundamental characteristic of life. They are not driven by the environment but originate from within — controlled by a biological clock carried in our genes. These daily fluctuations also occur in animals kept experimentally in continuous darkness or light.

The rhythms originate from the biological clock, but are constantly checked against the solar clock — the biological clock has to be reset daily. The most important external synchroniser is the daily light–dark cycle. This resetting of the biological clock means that the animal or plant is tuned to its environment and ensures that the right thing happens at the right time.

How biological clocks work

Biological clocks appear to be present in all living things. The mechanisms of the genetic clocks vary but even organisms as far apart as unicellular algae, fruit flies and mammals show similarities in how their clocks work. The differences may reflect their life history and the environment to which they are adapted.

Apparently every cell in every organism has a clock, but in vertebrates, a group of neurones in the central nervous system form a 'master clock'. It transmits timed signals that control the behaviour of the organism and adjust the many cellular clocks such that heart, kidney and liver function is coordinated with respect to environmental conditions.

Biological clocks never run to exactly 24 hours. This is why the patterns of behaviour and function observed when the clock is unsynchronised are termed 'circadian' (from

circa — about, dies — a day). They become synchronised by the light–dark cycle to the exact period of the solar day. In mammals, light input is via the eyes and our retinas have some special cells, different from rods and cones, which signal light or darkness to the master clock.

Experiments on biological clocks

In a study on Svalbard some students lived isolated in remote cabins during mid-summer (continuous daylight). They carried a device that recorded their body temperature and also when they went to bed and got up. The results showed that their sleep–wake rhythms shifted progressively from day to day. The absence of a daily light–dark cycle meant that their sleep–wake pattern was driven only by their biological clock, which became free-running. The volunteers lost synchronisation with the solar day even though they were aware of signs that tell the time, such as the sun circling the horizon and coming closer to the horizon each night. However, this mental information was apparently invisible to their biological clock. The experiment showed that in the absence of a clear day–night cycle, the brain master clock in humans issues signals based on its inherent circadian rhythmicity, that is, with intervals (periods) slightly longer than 24 hours.

Above the Arctic Circle the sun remains permanently below the horizon for some time in winter. Even though there is not permanent darkness, some people still experience a gradually shifting sleep–wake pattern. This is because neither the mid-day twilight nor the indoor lighting is bright enough to signal daylight to their biological clock. The same is true in Antarctica: during a winter expedition the participants became 'free-running', but people who stayed together tended to have similar rhythm lengths. Such social synchronisation is common and may explain why most people living above the Arctic Circle do not experience significant rhythm problems — they live by the pulse of modern society.

Migratory birds know the time

In order to synchronise their biological clock, people generally need to see a more marked light–dark cycle, i.e. brighter daylight, than most other organisms. When the activity patterns were recorded in captive birds during summer at Svalbard, it revealed a conspicuous resting period around midnight. This means that the birds' biological clocks were not free-running and they were able to perceive the synchronising information from the environment. Snow buntings, which visit every summer, and greenfinches, brought north for the occasion, showed similar patterns. It is unknown what signals they pick up, but birds may synchronise their activity patterns to the daily changes in light coloration resulting from the sun approaching the horizon each night (more red).

Another study showed that guillemot chicks all chose the same time to jump off the cliff nests where they had been hatched, to join their parents at sea. Most chicks made the risky jump just before midnight. This is a good strategy because there is safety in numbers. It reduces the chance of being eaten by foxes and gulls. Such coordinated behaviour is most likely caused by biological timing — their clocks would have to be synchronised with the environment for this to happen.

Reindeer and ptarmigan ditch their clocks

Humans and migratory birds are both visitors to the high Arctic. The Svalbard ptarmigan, however, and Svalbard reindeer are representatives of the few species that have lived there all year round for many generations. Experiments were carried out on both species to see if their biological clocks had adapted in any way to the extreme light–dark conditions.

Svalbard ptarmigan were kept in cages in the high Arctic and exposed to natural light and temperature conditions. Their feeding was monitored by photosensors. The results showed that whenever there was a light–dark cycle, the birds were mainly active during the day but during summer and winter there was no rhythm at all. During the polar day and night they were active around the clock with only short, scattered periods of rest.

In Svalbard reindeer, activity was recorded using data loggers attached to a neck collar while they were moving freely and undisturbed. The loggers responded to movement and stored values every 10–15 minutes. The results showed a pattern of activity in winter and summer similar to that adopted by ptarmigan. However, in spring and autumn, when there were marked light–dark cycles, reindeer showed only a weak reduction in activity during the night. Reindeer are ruminants and feed whenever they can.

This pattern was different from that among the reindeer living at lower latitudes in northern Norway. The 'southern' reindeer showed short but marked rest periods at dawn and dusk. In summer, when there was midnight sun, they were also permanently active. The different activity patterns in these two subspecies of reindeer may be explained by the different sensitivity of their biological clock to light, but may also relate to their different social organisation. The Norwegian reindeer are gregarious and form large herds with strong social interactions, while Svalbard reindeer are more solitary, forming only small groups of three to five animals. In addition, there are no reindeer predators on Svalbard.

Adaptations to Arctic light

The concentration of a hormone called melatonin was measured in Norwegian reindeer and Svalbard ptarmigan kept in Tromsø, a Norwegian town well above the Arctic Circle. As expected, there were marked daily rhythms in spring and autumn reflecting the duration of the night. In summer, however, melatonin concentrations were permanently low and undetectable, as if this important component of biological timing had stopped working. In people living in Tromsø there is a marked

night-time peak in melatonin throughout the year, coincident with sleep. However, in summer the peak is markedly delayed, presumably caused by people's tendency to stay outdoors in the evenings at this time of year. It is not known if and how reindeer, ptarmigan and humans living under Arctic light condition are affected by their melatonin profiles. However, it is likely that the changes in melatonin secretion reflect changes in their biological clock.

Studies at Svalbard have revealed three different patterns of behaviour throughout the polar day. Humans are not adapted to life at these latitudes and in some ways could be regarded as visitors from the tropics. In the absence of rhythmic input from society and a marked light–dark cycle, we are driven by our inner biological clock. Migratory birds spend most of the year at lower latitudes and, like humans, are visitors to Svalbard. The birds, however, have been subjected to strong natural selection. Their biological timing mechanisms are apparently highly sensitive to temporal signals during the polar day. It may be important for them to display synchronised behaviour, for example the jumping of guillemot chicks.

The most remarkable pattern is seen among the native reindeer and ptarmigan. For both species it appears as if their biological clock is not running, or is only weakly expressed. Moreover, Svalbard reindeer are virtually unaffected by the marked light–dark cycles in spring and autumn. We may explain this as an adaptation to living in an environment where there is little difference between night and day for most of the year. Thus, the polar form of biological timing is to reduce the input from a biological clock. In summer and winter, when time 'stands still', it may be unfavourable to have an endogenous clock that drives the organism through subjective days and nights. It may be wiser to live opportunistically and always be prepared to act and feed. Arctic animals avoid being in 'night-mode' when weather conditions allow them to feed — it may be a long time until it happens again.

Source: Biological Sciences Review, *November 2008*

Scientific article 2
Refractory period

Refractory period is a term used in connection with nerve and muscle cells. During the refractory period these excitable cells become unable to respond to stimuli that would normally cause them to generate action potentials. In this What Is?, we look at what the term means, what happens at the plasma membrane during the refractory period, and how the refractory period contributes to the normal and abnormal functioning of these cells.

Dictionary definitions of 'refractory' include 'hard to manage', 'disobedient' and 'resisting ordinary methods of treatment'. These descriptions imply that a refractory period is a time during which what we might consider to be the norm does not occur, and that is exactly the case. Both nerve and muscle cells generate electrical impulses (action potentials), which have a specific duration. This is usually a few milliseconds for a nerve cell but may be much longer — more than 100 ms — in a heart muscle cell. The generation of an action potential is triggered by depolarisation of the plasma membrane. This makes the cell less negative inside, up to or beyond a certain threshold level. Once the threshold is reached, an action potential is always generated. An exception to this is during the refractory period — a short period of time immediately after an action potential, during which it is much harder to generate another action potential. The refractory period is divided into two parts. At first, no action potential can be triggered at all — this is called the absolute refractory period. The relative refractory period then follows, during which an action potential may be triggered, but only if a greater than normal depolarising stimulus is supplied, that is, the threshold is raised.

How does it work?

To understand how the refractory period works, we need to review briefly the events that occur at the plasma membrane during an action potential. When a nerve cell is at rest, its plasma membrane is polarised, that is, it is negatively charged inside compared with outside the cell. The resting membrane potential is usually around −70 millivolts. When the membrane is depolarised beyond a threshold voltage (around −50 mV), sodium ion (Na^+) channels in the membrane open. These channels allow Na^+ to diffuse rapidly into the cell, where the concentration of Na^+ is much lower than in the extracellular fluid. Influx of Na^+ causes further depolarisation because of the positive charge of the sodium ions. It is this positive feedback cycle of depolarisation → opening of Na^+ channels → further depolarisation that leads to the fast-rising phase of the action potential. However, Na^+ channels have a strictly time-limited opening. After around 0.5 ms, they automatically shut and the voltage across the membrane begins to return to its resting level of −70 mV. During the falling

phase of the action potential the Na⁺ channels are in an 'inactivated' state and they cannot open again until they return to their normal 'closed' state. The time during which Na⁺ channels are inactivated represents the absolute refractory period; no depolarisation, however large, can trigger a further action potential.

Potassium ion (K⁺) channels also contribute to the falling phase of the action potential. These open later than Na⁺ channels, during the falling phase. When open, they allow K⁺ ions to leave the cell, making it more negative inside and thus aiding the return to the normal resting potential of –70 mV. However, K⁺ channels stay open for longer than Na⁺ channels and this makes the membrane potential briefly overshoot to a level more negative than at rest. This overshoot is called the after-hyperpolarisation and follows every action potential. During the time when K⁺ channels are open, it is much harder to depolarise the cell to threshold and this represents the relative refractory period. The action potential and refractory period in heart muscle cells work in a similar way, but last much longer.

Why is it needed?

The refractory period ensures that action potentials are spaced out, rather than overlapping each other. It is easy to see how important this is for heart muscle action potentials. Since each impulse in the heart muscle cells triggers muscle contraction, it is essential that beats are adequately spaced so that there is time for the heart to fill up with blood in between.

In nerve cells, spacing between action potentials over time allows for the 'digital' coding of information. Each nerve fibre (axon) has a maximum firing frequency, which is determined by the duration of its refractory period — the longer the refractory period, the slower the maximum firing rate. So, for example, sensory axons in your auditory nerve may fire up to 500 action potentials per second — this is because their refractory period is only about 1ms long. In contrast, the maximum firing rate of some nerve cells in the brains of slugs and snails may be as little as 20 action potentials per second. Below the maximum frequency, the strength of a stimulus is coded for by the frequency of action potentials passing along the axon. This is called the 'frequency code'. Sensory nerves, as a general rule, code for increased intensity of a stimulus (light, sound, touch) by increasing their action potential firing frequency.

The refractory period also ensures that action potentials only travel in one direction along an axon. Imagine an action potential that is triggered half-way along an axon. There is no reason why it should not travel in both directions away from the site of initiation. However, in the normal situation where an action potential travels from the cell body towards the axon terminal, the portion of axon just behind the action potential is in a refractory state because it has only just finished generating an action potential. Therefore the action potential can only travel one way, that is, away from where it has just come.

questions & answers

Can it go wrong?

Abnormalities in nerve cell refractory period occur in conditions such as diabetic and alcoholic neuropathies (diseases of nerves). These conditions can be extremely debilitating, with a huge variety of symptoms such as severe pain, numbness, itching, tremor, muscle weakness and abnormalities in posture and movement. The refractory period is often prolonged, impairing the ability of the nerves to conduct action potentials at high frequencies. Similarly, in diseases where there is loss of the insulating myelin sheath around the axons — for example, multiple sclerosis — the refractory period is again abnormally long. This can severely restrict the maximum frequency at which action potentials can be transmitted along axons. A common early symptom of multiple sclerosis is a reduction in the ability to sense vibration. This is because accurate coding of vibrating stimuli on the skin is highly dependent on the sensory axons' ability to transmit action potentials at high frequency. Later stages of the disease involve impairment in both sensory perception and muscle control, as both sensory and motor axons lose their ability to transmit action potentials.

A normal refractory period is essential in heart muscle cells to avoid the heart going into a state of fibrillation. Fibrillation in the ventricles of the heart (ventricular fibrillation) can occur when contraction of heart muscle cells becomes uncoordinated. Because the heart can no longer pump blood efficiently, the body quickly becomes deprived of oxygen, leading to unconsciousness and cardiac arrest. This condition is often fatal and can occur when the refractory period is, for whatever reason, abnormally short.

In the normal situation, an action potential travels quickly through all the muscle cells of the ventricle for every heart beat, leading to coordinated contraction to pump blood out. The cells are all then in a refractory state and thus cannot contract again until the next heartbeat is triggered. The normal refractory period in ventricular muscle cells is around 300 ms (about one-third of a second). However, if the refractory period becomes much shorter, then by the time the wave of activity has passed through the whole ventricle, many cells are no longer refractory and can become active again. The whole ventricle can enter a state in which individual muscle cells generate action potentials and contract at high frequency and in an uncoordinated and disorganised fashion. The heart 'quivers' and normal pumping of blood ceases. Treatment is by 'defibrillation'; a large electric shock is applied to the heart, making all the cells fire at once and then become simultaneously refractory. The hope is that normal function will resume within a few seconds and the patient can then be treated with drugs to prolong the refractory period.

Source: Biological Sciences Review, *April 2008*

scientific article

The cerebral cortex

The cerebral cortex is only a few millimetres thick, yet it plays a vital role in our interaction with the world. The cortex receives, analyses and stores information from our senses and initiates voluntary movement. Behaviourally sophisticated mammals, such as humans, have larger cortices than lower mammals. This reflects the complexity of the coordination of the tasks they perform. The cortex has specialised regions, each carrying out a particular function. The frontal lobes process thought and play a role in memory retrieval. The senses also have their own regions. Hearing is processed in the auditory cortex, which is in the temporal lobe; vision resides in the occipital lobe. The parietal lobe contains the motor and somatosensory cortices, in which voluntary movement and the sense of touch lie, respectively. All these areas are found in both the right and left hemispheres. Other functions are localised in just one hemisphere, such as the language areas, found on the left.

The cortex both receives information from our surroundings and sends commands to our muscles. Sensory receptors, such as touch receptors in the skin, are part of the peripheral nervous system, and respond to environmental stimuli. Impulses are then sent to the relevant sensory cortex through peripheral nerves. The area of cortex the impulses reach first is called a primary cortex. Here, crude features of the stimulus are decoded, such as the orientation of a visual stimulus in the primary visual cortex.

The stimulus is recognised by association areas that compare it to previous knowledge. Each of the senses has its own primary and association area. The association areas communicate with many brain areas, including the frontal lobe and memory- and emotion-associated regions. The frontal lobe also receives input from a number of other regions before deciding on action. The premotor cortex (also called the motor association area) coordinates and plans movement, while the primary cortex sends nervous impulses to our muscles in order to initiate contraction.

Principles of cortical organisation

A cortical map shows the size and location of cortical areas. Cortical mapping dates back to the 1930s when the cortex was electrically stimulated to reveal which body parts it governed. Mapping now uses scanning methods, such as functional magnetic resonance imaging (fMRI). This technique detects areas of the brain that are active during a given task, by detecting brain regions with an increased blood flow.

In cortical maps, sensory surfaces that are behaviourally important occupy the most space. For example, hands are vital for human communication and survival. Although they cover a smaller skin area than the abdomen, the impulses from them occupy a much larger space in the somatosensory cortex. This reflects differences in the density of skin sensory receptors. Sensitive areas, such as the hands, have a high concentration of sensory receptors and therefore require larger cortical areas for processing. Similarly, body parts requiring fine motor control have large representations in the motor cortex. Thus the size of a cortical area relates to the number of sensory receptors, or muscle effectors, in the corresponding skin or muscle area.

Cortical representations of our senses and muscles are not fixed. What makes the cerebral cortex so fascinating is its capacity to change the distribution of different functions. Cortical space can be reallocated to govern or respond to different parts of the body, either as a response to injury or to learning. This capacity to reorganise is termed cortical plasticity, and is most effective in childhood.

Cortical adaptation to injury

The nervous system can compensate for damage in remarkable ways. For example, what happens when a finger is amputated? Does the cortical area corresponding to the finger stay devoted to the lost finger? The answer is 'No'. As the sensory input from the amputated digit is lost, the corresponding cortical area starts receiving input from other skin areas, such as the neighbouring digits and adjacent palm areas. By employing this trick, the cortex maximises the use of cortical space. This happens not only at the level of individual skin regions/maps, but also on a wider scale. Blind people process some tactile information in the occipital lobe, which is normally associated with vision. Some of the visual processing of deaf people takes place in the temporal lobe, which is normally associated with hearing. This shift in processing from one cortical area to another highlights the brain's ability to respond to injury.

Musical learning reshapes the cortex

Since the cortex is involved in processing new experiences, how does it learn from them? Learning is a process that affects many parts of the brain, especially the cortex. Playing a musical instrument is an example of a task that involves many cortical regions. How does the cortex respond to years of practice ultimately culminating in increased skill? An answer to this question was found in the brains of violinists. When comparing somatosensory maps of violinists' fingers in the two hemispheres, it was found that their left-hand digit areas, which lie in the right somatosensory cortex, were enlarged. This enlargement can be explained by the left-hand fingers being responsible for pressing the strings, while the right hand handles the bow. Thus, the left fingers require a higher level of fine motor coordination. This finding highlights an important point about cortical plasticity: the more stimulation an area receives, the larger its representation in the sensory cortex.

But, if cortical representations keep on expanding with experience, why doesn't the cortex eventually run out of space? For this the brain has developed a coping strategy. Space is reallocated to regions of importance, at the expense of other neighbouring regions, which diminish in size. In the case of violin players, the finger areas expand into the adjacent palm region, which receives less cortical space. The reverse is true for rarely stimulated skin areas; their cortical areas tend to diminish.

Musical training also affects motor maps. It takes merely a few days of piano practice to enlarge finger areas in the primary motor cortex of adults with no previous musical training. This enlargement parallels the changes observed in the somatosensory system.

Communication across the cortex

Practising music not only causes individual cortical areas to rearrange, it also induces cortical areas to 'talk' to each other. Have you ever had a tune stuck in your head that is impossible to silence? For this you can blame your auditory cortex, which is remarkable in that it can activate in the absence of sensory input. Playing a song internally activates the auditory cortex, resulting in the auditory experience. What fuels this illusion of 'hearing'? It seems that frontal lobes also activate during this process. Remember that these lobes are involved in memory retrieval, and thus serve as a 'song-feeder' to the auditory cortex.

Another example of intercortical communication involves the motor cortex. The auditory and visual cortices help time and coordinate movements when rehearsing a piece of music. When a pianist listens to, or sees, a well-rehearsed piece being performed, the motor cortex is activated. This activity does not cause finger muscle contractions, and thus signals intention rather than actual movement. Specialised neurones, called mirror neurones, fire during this process. These cells are thought to underlie our capability to imitate observed movements and are therefore important in learning.

Cortical plasticity for better and for worse

Cortical plasticity comes with a price, however. While clearly important in learning, the cortex is limited in terms of how much instrumental training it can cope with. Musicians' cramps (focal dystonias) involve loss of individual finger movement control and can be devastating for professional musicians. This condition particularly affects pianists and was until recently treated with drugs. Dystonias are caused by fusion and overlapping of individual digit areas in the somatosensory cortex. This discovery enabled the development of training schemes involving finger movement, and touch discrimination tasks to replace drug treatment. This kind of training successfully recovers normal hand functioning. So, cortical plasticity can be applied to reverse some of the negative changes in the cortex caused by excessive practice.

Source: Biological Sciences Review, *September 2008*

Question 1

One aspect of vision is called visual acuity. This is the ability to distinguish small objects as separate from one another. It is often tested by showing subjects a plain grey object and an identical one in which narrow lines are drawn as stripes or a grid. The narrower the lines that can be distinguished from the grey object, the greater the visual acuity.

(a) Small children are not able to tell us what they see, but we do know that they spend more time looking at patterned objects than uniform ones. Suggest how this fact could be used to determine the visual acuity of a one-week-old baby. (3 marks)

(b) The retina contains two types of receptors, rods and cones. Cones are tightly packed in the area onto which light is focused when looking directly at an object. Rods are spaced more widely apart. Suggest which of these types of receptor cells are responsible for visual acuity, and explain your suggestion. (2 marks)

(c) The graph shows the change in visual acuity during the first 2 years after birth.

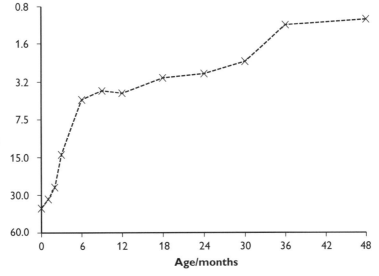

(i) Describe the change in visual acuity between 0 and 2 years of age. (2 marks)

(ii) Babies are sometimes born with cataracts, which prevent light from entering the eye and being focused on the retina in the normal way. It is found that, following the removal of the cataracts at 12 months old, these babies have a visual acuity that is less than that of babies who had fully functioning eyes from birth, and their visual acuity is still less than normal when they reach adulthood.

To what extent does this provide evidence for the existence of a critical 'window' in the first 12 months of life for the development of visual capacity?
(4 marks)
Total: 11 marks

■ ■ ■

Candidate A

(a) You could hold two pieces of card up, one just grey and one with stripes. See which one the baby looks at most. ✓ Keep doing that with different sized stripes. If it keeps looking at the stripes more than the grey card then it can see the stripes. ✓

> Quite a good answer — the candidate has described a suitable procedure. However, more detail is needed to explain how the visual acuity would be measured. 2/3

(b) Cones, because they are the ones that are most tightly packed ✓ so they could pick up more detail ✓ about an image.

> A correct choice, and just enough in the explanation for both marks. 2/2

(c) (i) The visual acuity decreases ✗ from just above 30 arbitrary units to about 1.2 arbitrary units, that is by about 28.8 units. ✓ It gets worse ✗ as the baby gets older.

> The student has misunderstood what is meant by 'visual acuity', or what is shown on the graph. The smaller the objects that can be distinguished, the better the visual acuity. However, he or she has correctly read off the start and end values and has calculated the change over the 2 year period. 1/2

(ii) This shows that visual acuity needs to have light coming into the eyes ✓ so that it can develop during the first 12 months. So this implies that the first 12 months are a critical period. ✓ If the babies can't see normally then, their visual acuity never develops properly even when they are adults. ✓

> A clear answer, but the candidate has not appreciated that the critical window could extend beyond the first 12 months, as we have no information about what happens if the cataracts are removed at a different time. 3/4

Candidate B

(a) Place the baby so that he/she is sitting comfortably. Keep the light intensity and colour constant. ✓ Hold up two cards the same distance ✓ from the baby. One card is plain grey and the other one has stripes a measured distance apart. ✓ Note which card the baby looks at longest. ✓ Repeat using the same grey card but with stripes less distance apart. Keep doing this until the baby looks at both cards for the same amount of time. The distance of the stripes on the previous card to this one is the limit the baby can tell apart. ✓

📝 An excellent answer, paying full attention to the control of variables and how the visual acuity would be determined. 3/3

(b) Cones are responsible for visual acuity. Being closely spaced, ✓ they can pick up more individual 'bits' of information, ✓ e.g. from closely-spaced objects.

📝 A correct answer. 2/2

(c) (i) Visual acuity increased throughout the time period. ✓ The greatest rate of increase was during the first 9 months, ✓ then there was slower increase from 9 to 30 months, then a faster increase up to 36 months. After this, it increased only a little. ✓

📝 A good answer, describing the main areas of the graph where the gradient changes. Although no calculation has been done, this is still enough for full marks. 2/2

(ii) This clearly shows that there must be a critical window somewhere within the first 12 months, ✓ because if the baby's eyes don't get normal stimulation during this time ✓ then it never develops its visual acuity completely. ✓ We cannot tell from this information exactly when this window is — it could be just a short period within that 12 months, ✓ or it could even extend beyond 12 months. ✓

📝 A full and entirely correct answer. 4/4

Question 2

(a) **Describe the structural and physiological differences between fast
twitch and slow twitch muscle fibres.** (4 marks)

(b) **Suggest how you would expect the relative proportions of these two
types of fibres to differ in a world-class sprinter and a world-class
distance runner. Explain the reasons for your suggestion.** (3 marks)

Total: 7 marks

■ ■ ■

Candidate A

(a) Fast twitch fibres are bigger than slow twitch fibres and they can contract more
quickly. They don't have as many mitochondria ✓ and they are not as red because
they do not contain as much haemoglobin. ✗

> This candidate has the right ideas, but has expressed the answer poorly and has
> made one error. The word 'bigger' is not clear enough; the candidate should have
> said that fast twitch fibres are 'wider' than slow twitch fibres. The pigment that is
> found in these muscle fibres is myoglobin (an oxygen storage compound) not
> haemoglobin. The statement that fast twitch fibres contract more quickly is correct
> but does not go quite far enough for a mark. 1/4

(b) A sprinter would have a lot of fast twitch fibres compared to slow twitch fibres,
and a long distance runner would be the opposite. ✓ This would mean the sprinter
could run faster than the long distance runner.

> One correct statement, but more explanation is needed. 1/3

Candidate B
(a)

Fast twitch	Slow twitch
Use mostly anaerobic respiration	Use mostly aerobic respiration ✓
Do not have many mitochondria	Have many mitochondria ✓ so they can carry out Krebs cycle and oxidative phosphorylation to produce ATP ✓
Are quite wide, as they don't need an oxygen supply	Are narrower, so oxygen can diffuse into them quickly ✓
Don't have much myoglobin because they don't need oxygen	Have a lot of myoglobin ✓ so they have an oxygen store in case the blood doesn't bring them enough oxygen
Few capillaries supplying them	Are supplied with oxygenated blood by many capillaries

It is a good idea to answer questions involving comparisons as a table, as it makes sure that you give both sides of the argument. This is a very good answer. 4/4

(b) The sprinter would have a greater proportion of fast twitch fibres compared to the distance runner. ✓ Fast twitch fibres can contract quickly, but can only keep it up for a short time ✓ because they respire anaerobically, but this is fine in a short sprint race. ✓ Slow twitch fibres can keep going for longer, which is what is needed for a distance runner, ✓ but they aren't much good to a sprinter because they contract too slowly.

A thorough answer, well expressed. 3/3

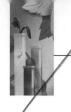

Question 3

The rate of aerobic respiration of yeast cells can be measured using alkaline phenolphthalein indicator. This changes colour from purple to colourless when the pH of a solution falls to a certain level.

A student set up four test tubes. Each tube contained the same volumes of yeast suspension, glucose solution and alkaline phenolphthalein indicator. Ethanol was also added to three of the tubes, to produce ethanol concentrations as shown in the table. The table also shows the time taken for the indicator in the tube to become colourless.

Tube	Ethanol concentration (%)	Time taken to become colourless/s
1	0	32 *more acid*
2	1.5	96
3	12.5	193
4	25.0	Did not become colourless *acid*

(a) Explain why the indicator lost its colour in tubes 1, 2 and 3. (3 marks)

(b) (i) Describe the effect of ethanol on the rate of respiration in yeast. (2 marks)

(ii) Suggest reasons for the effect you described in (i). (2 marks)

(c) State two variables that should be kept constant during this experiment, and explain how the student could achieve this. (4 marks)

Total: 11 marks

■ ■ ■

Candidate A

(a) This is because the yeast made carbon dioxide, ✓ which makes an acid when it dissolves in water.

🖻 This is a correct answer, but the explanation is not full enough for a 3 mark question. 1/3

(b) (i) The ethanol inhibited respiration. ✓ At a concentration of 25% it stopped it completely. ✓

🖻 A correct general point and then a specific one. 2/2

(ii) Perhaps the alcohol is toxic 3 to yeast.

🖻 A reasonable suggestion, but more is needed for a second mark. 1/2

(c) Temperature. ✓ Put it into a water bath. ✓ Concentration of glucose solution. ✓ Measure all the glucose from the same solution. ✓

🖉 Two relevant variables to be controlled, and outlines (though very brief) of how they would be controlled. 4/4

Candidate B

(a) As the yeast respired, ✓ it produced carbon dioxide, ✓ which dissolved in the water and made carbonic acid. ✓ This is a weak acid which lowers the pH. ✓ The more carbon dioxide, the lower the pH, until it goes below the point at which alkaline phenolphthalein changes from purple to colourless. ✓

🖉 A very good explanation. 3/3

(b) (i) The greater the concentration of alcohol, the slower the rate of respiration. ✓ It seems that respiration did not happen at all when there was 25% ethanol, ✓ but we cannot be sure it might not have eventually changed colour if it had gone on longer.

🖉 A good answer. 2/2

(ii) Perhaps ethanol inhibits one of the enzymes ✓ involved in respiration, by binding with it at a point other than its active site and changing the shape of its active site. ✓ Or it might be a competitive inhibitor, competing for the active site with the enzyme's normal substrate. ✓

🖉 A good suggestion, expanded to give two alternative ways this could happen. 2/2

(c) Temperature; place all the apparatus in a thermostatically-controlled water bath at about 30°C. Concentration of yeast; make up a large quantity of yeast suspension, and stir it thoroughly before measuring out the same volume into each tube.

🖉 Two suitable suggestions, and the methods of controlling them well explained. 4/4

Question 4

(a) With reference to temperature regulation in humans, explain the meaning of each of the following terms.

    (i) negative feedback   **(3 marks)**

    (ii) dynamic equilibrium   **(3 marks)**

(b) Explain why core body temperature tends to rise during strenuous exercise.   **(2 marks)**

(c) Ventilation rate increases during exercise. Describe how this increase in rate is brought about.   **(4 marks)**

Total: 12 marks

■ ■ ■

Candidate A

(a) (i) This is when something changes in one direction and then the body does something to make it change back again. ✓

    This is correct, but it is not a very clear explanation and it does not refer to temperature regulation in humans, as the question asked. 1/3

    (ii) This is when things don't settle down and stay at one level, but keep changing a bit all the time. ✓

    Again, the candidate has the right idea, but the explanation is not clear and again it does not refer to humans. Just enough for 1 mark. 1/3

(b) The muscles are working hard and so they get hot.

    Once more, there is nothing incorrect in this answer, but it is not sufficient for an A level explanation. 0/2

(c) When you are exercising a lot, there is more carbon dioxide in the blood and this is detected by chemoreceptors. ✓ They send messages to the brain which responds by making the breathing muscles work faster.

    Again, lacking in the detail expected in an A level answer. The term 'messages' is not a good one to use — it is best to use correct scientific terms such as 'action potentials' or 'electrical impulses'. 1/4

Candidate B

(a) (i) In a negative feedback loop, a receptor detects a change in body temperature, and this causes an effector to bring about changes ✓ in the oppposite direction. ✓ For example, if core temperature rises then the hypothalamus detects this and stimulates effectors to do things that bring the temperature down. ✓

🖉 It isn't easy to describe negative feedback briefly, and this candidate has done well. The answer also refers specifically to temperature regulation. 3/3

(ii) In negative feedback loops, there is always a delay between the receptor picking up information about the change and the effectors doing something ✓ to make it change the other way. So the temperature doesn't stay absolutely constant ✓ but keeps oscillating on either side of the 'correct' temperature. ✓

🖉 An entirely correct explanation, once again making reference as asked to temperature regulation. 3/3

(b) Some of the energy released from glucose during respiration ✓ in the muscles is released as heat instead of being used to make ATP. ✓ This increases the temperature of the blood, ✓ which transfers the heat to other parts of the body.

🖉 A good answer. 2/2

(c) Ventilation rate is the number of breaths per minute × the volume of each breath. ✓ Respiration ✓ in muscles produces carbon dioxide, which dissolves in the blood and lowers its pH. ✓ This is detected by receptors ✓ in the ventilation centre in the brain, and also in the aortic body ✓ and carotid bodies. This causes the ventilation centre to send more frequent nerve impulses ✓ to the diaphragm and intercostal muscles, so they contract more strongly and more frequently. ✓

🖉 An excellent and very complete answer. 4/4

Question 5

IgA is an immunoglobulin (antibody) that is particularly important in controlling infections in the tissues lining the respiratory passages. The graph shows how the quantity of IgA in the saliva of an elite kayaker changed during a 13-day intensive training period.

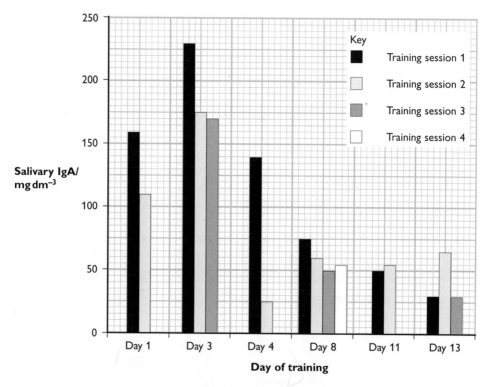

(a) (i) Calculate the percentage change in salivary IgA concentration between the first and second training sessions on Day 1. Show your working. (3 marks)

(ii) Describe the changes in salivary IgA during the 13-day training period. (4 marks)

(b) It has been known for many years that intensive training is often associated with an increased likelihood of suffering from an upper respiratory tract infection (for example, a cold). Discuss the extent to which the data in the graph provide evidence for a causal link between training and upper respiratory tract infections. (3 marks)

Total: 10 marks

Candidate A

(a) (i) $160 - 110 = 50$ ✓

$50/160 \times 100 = 31\%$ ✓

> 🖉 A correct calculation, but the answer does not state whether the change is positive or negative. 2/3

(ii) The concentration went down overall. ✓ The highest amount was after the first session on day 1, and the lowest was after the first session on day 13. However, on day 2 it went up ✓ for some reason, and was higher after all the training sessions than it was at any point on day 1. On days 1, 3, 4 and 8, it was lower after the second training session than after the first training session, ✓ but on days 11 and 13 the opposite happened. ✓

> 🖉 The candidate has picked up the most significant trends on the graph. 4/4

(b) IgA helps to fight off viruses that might give you a cold, so it makes sense that if there is less IgA then you stand more chance of getting a cold. ✓ So this does suggest there is a link between getting a cold and training.

> 🖉 The candidate has not fully focused on the issue of whether this particular set of data indicates a causal link between training and getting a cold. 1/3

Candidate B

(a) (i) Difference between first and second sessions $= 160 - 110 = 50$ ✓ mg dm^{-3}

So percentage change $= (50 \div 160) \times 100$ ✓ $= -31\%$ ✓

> 🖉 Entirely correct, including showing that the percentage change is a decrease. 3/3

(ii) The concentration went down after the second training session on each of the first 8 days for which we have information, ✓ but after that it went up. It also went up between day 1 and day 2, ✓ but after that it kept going down each day. ✓ The biggest change between training sessions was on day 4, ✓ when it went down about 80% ✓ after the second training session.

> 🖉 The main trends and patterns have been picked out, and a calculation has been made relating to one of the significant changes. 4/4

(b) The graph does not tell us anything about infections, only about the levels of IgA. ✓ So the data could support the hypothesis that training causes more respiratory infections, because it would make sense that if there is less IgA then you have a bigger chance of getting infections. ✓ But the data don't actually prove this, so we would have to measure these same things again in lots of different people and also measure the number of colds they got. ✓ Even then we wouldn't be absolutely sure there was a causal link, because there might be some other factor we haven't measured that is causing the difference. ✓

> 🖉 Some good points have been made here — the first one is especially important. 3/3

Question 6

The diagram shows a vertical section through a human heart.

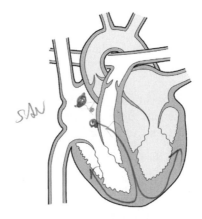

SAN

(a) On the diagram, label:

   (i)   the position of the sinoatrial node (SAN)   (1 mark)

   (ii) the position of the atrioventricular node (AVN)   (1 mark)

   (iii) the path taken by the wave of electrical activity during one heart
         beat.   (2 marks)

(b) Heart muscle tends to use fatty acids as the main respiratory substrate,
    rather than glucose. Fatty acids are first converted to acetyl coenzyme A
    before taking part in the metabolic pathways of respiration.

   (i)   Explain how this information helps to explain why heart muscle is
         unable to respire anaerobically to any great extent.   (3 marks)

   (ii) Outline how reduced NAD is used to produce ATP in the inner
         mitochondrial membranes of heart muscle.   (4 marks)

                                                              Total: 11 marks

■ ■ ■

Candidate A

(a)

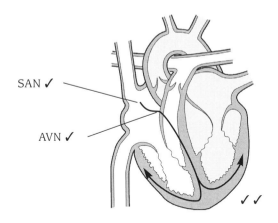

SAN ✓

AVN ✓

✓ ✓

🖉 All correct. 4/4

(b) (i) It couldn't do glycolysis if it doesn't use glucose. ✓

🖉 There is the beginning of a correct answer here, but much more needs to be included in the explanation. 1/3

(ii) The reduced NAD gives its electrons ✓ to the first thing in the electron transport chain. ✓ The electron goes along the chain and is used to make ATP. This happens inside the mitochondria and it is called oxidative photophosphorylation. ✗

🖉 The answer begins correctly, but we are not told any detail about how the ATP is made. The term 'photophosphorylation' is incorrect. 2/4

Candidate B

(a)

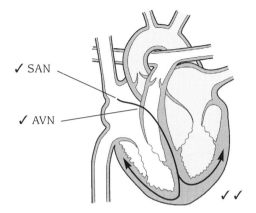

✓ SAN

✓ AVN

✓ ✓

🖉 All correct. 4/4

(b) (i) The fatty acids will have to go into the mitochondria and when they turn into acetyl coenzyme A they will go through Krebs cycle. ✓ This can only happen when there is oxygen available. ✓ Anaerobic respiration can only be done using glycolysis ✓ as a starting point, which uses glucose and changes it to pyruvate which is then converted to lactate. ✓ Fatty acids can't be used for glycolysis.

A good answer. 3/3

(ii) Hydrogens from the reduced NADP are released ✓ and split into protons and electrons. The electrons go along the carriers ✓ in the inner mitochondrial membrane, losing energy ✓ as they are passed along. The energy is used to pump hydrogen ions into the space between the membranes. ✓ When the hydrogens diffuse back out, they go through ATPases ✓ which use the energy to combine ADP and P_i to make ATP. This is called chemiosmosis. ✓

Entirely correct. 4/4

Question 7

The scientific document you have studied (pages 47–54) is adapted from articles on synapses and biological clocks in *Biological Sciences Review*. Use the information in the document and your own knowledge to answer the following questions.

(a) Describe what is meant by a ligand-gated channel, and outline the usual roles of ligand-gated channels at a synapse where the neurotransmitter is acetylcholine. (4 marks)

(b) Nicotine is not normally found in the body, but some synapses are described as having 'nicotinic receptors'.

   (i)   Explain why the receptors at these synapses are given this name. (2 marks)

   (ii) Explain how and why depolarising blockers are used in surgery. (3 marks)

(c) Glutamate is a neurotransmitter found in the brain. Monosodium glutamate is a commonly used flavour enhancer and there has been some concern that very high levels of monosodium glutamate in the diet might cause brain damage in young children.

   (i)   Suggest why high concentrations of glutamate in the diet might cause brain damage. (3 marks)

   (ii) Discuss the arguments for and against using animals in research to determine the effects of high levels of dietary glutamate in young children. (4 marks)

(d) Melatonin is a hormone secreted by the pineal gland, which is involved in the regulation of circadian rhythms in humans, birds and other vertebrates.

   (i)   Explain the similarities and differences between neurotransmitters and hormones. (5 marks)

   (ii) Suggest how plants may detect changes in light intensity over a 24-hour period. (2 marks)

   (iii) With reference to specific examples, discuss the extent to which circadian rhythms in living organisms are controlled by genes and by environment. (7 marks)

Total: 30 marks

■ ■ ■

Candidate A

(a) Ligand-gated channels are the body's molecular switches. They usually let sodium ions in through the post-synaptic membrane. ✓

> The candidate has just copied out a sentence from the article to begin the answer. This is not usually a good thing to do, because the person who wrote the article was not trying to answer this question! You need to use the information in the article, and your own knowledge, to construct an appropriate answer to the question. 1/4

(b) (i) They can bind with nicotine. ✓

> Just enough for one mark. 1/2

(ii) They are used to paralyse the patient so that they do not move around during surgery ✓ which could make the surgeon make mistakes and cut the wrong thing. They work because they stop acetylcholine binding with its receptors at synapses. ✓

> All correct, but more needed. 2/3

(c) (i) They might damage nerve cells in the brain.

> This is not a good enough answer to earn any marks. 0/3

(ii) It wouldn't be morally or ethically acceptable to do experiments on young children so you need to do them on animals instead. ✓

> This is a very relevant point, but the question asked the candidate to 'discuss', so two different points of view need to be included in the answer. 1/4

(d) (i) They are both chemicals. They both make things happen in the body. Neurotransmitters are made by neurones and hormones are made in endocrine glands. ✓

> This is quite a weak answer. This is not an easy question, and the answers will not be found in the article, although some of the passages in the article describing how neurotransmitters act could give you ideas about linking this to what you know about how hormones act. 1/5

(ii) They have photoreceptors called cryptochromes, phototropins and phytochromes. ✓

> This list of photoreceptors is correct, but more is needed to answer this question. 1/2

(iii) We have a 'biological clock' in the body. The mechanism of the genetic clocks vary, but even organisms as far apart as unicellular algae, fruit flies and mammals show similarities in how their clocks work. Every cell has a clock, and there is also a 'master clock' in the brain. But the clock can be adjusted depending on the light-dark cycle. ✓ For example, some students lived in a place where there was constant daylight and their clock started to run on a cycle longer than 24 hours. ✓ Ptarmigans were active in the day when there was a light-dark cycle, but when it was completely light or completely dark they don't have a rhythm of activity. ✓ This shows that their genetic clock doesn't really do anything on its own. It needs light-dark cycles to make it work. ✓

🖉 This is a difficult question, requiring not only selecting relevant information from a long article, but also manipulating it to answer a question which is not addressed in the article itself. This candidate has found it hard, but neverthless has managed to make some relevant points. Note that the second sentence of the answer is copied directly from the article; this is never a good idea. Marks have been given for the idea that there is an underlying genetic clock that is modified by light-dark cycles, with some contrasting examples — the students whose internal clock continued to run with a rhythm even during constant light, and the ptarmigan whose internal clock seemed to require light-dark cycles to run at all. 4/7

Candidate B

(a) A ligand is a molecule that binds with a receptor. ✓ So, for example, acetylcholine is a ligand which binds with receptors in the post-synaptic membrane. ✓ If the binding of the ligand causes a channel to open in a membrane, ✓ this is called a ligand-gated channel. So when acetylcholine binds with its receptors, it causes the channels to open and allow Na^+ ✓ to flood into the neurone down their electro-chemical gradient. This depolarises ✓ the post-synaptic neurone and may cause an action potential in it.

🖉 A good answer. The candidate has used information about ligand-gated channels from the article, and then described how one works using his or her own knowledge about synaptic transmission. 4/4

(b) (i) These receptors normally bind with acetylcholine. However, some acetylcholine receptors will also bind with nicotine if it is present. ✓ This is just an accident — the body shouldn't have any nicotine in it so it is just chance that the shape of a nicotine molecule is similar to an acetylcholine molecule. ✓

🖉 A correct answer. 2/2

(ii) A depolarising blocker is something that will bind with the acetylcholine receptors on the post-synaptic membrane (the ones that nicotine can bind with) and make the sodium ion channels open. ✓ So the muscle cell gets depolarised. But the blocker doesn't get broken down so the muscle stops working. ✓ This is useful when surgeons are operating on someone, because it means the muscle stays really still. ✓

🖉 A well-constructed answer. First the candidate explains what a depolarising blocker is and how it works, and then links this to its use in surgery. None of this has been directly copied from the article — the candidate has used his or her own words throughout. 3/3

(c) (i) We know that after a stroke too much glutamate can be released ✓ and this can make too much calcium go into brain neurones ✓ and this can kill them. ✓ So if you eat a lot of glutamate the same thing might happen.

🖉 Again, a good answer. It begins with an explanation of how glutamate can damage neurones in the brain, which is well described. 3/3

(ii) The best way to find out how glutamate in the diet affects young children would be to give some children a lot of glutamate and others no glutamate and see how it affects their brains. But obviously we can't do that, ✓ so animals can be used instead. You can't really use tissue culture because tissues in tissue culture are just lots of cells all the same kind, which aren't the same as brains. ✓

🖉 This is not such a good answer as the previous ones. It gives only one side of the argument — why animals should be used for this research — but does not address the other side — reasons why we should not use animals in this research. 2/4

(d) (i) Neurotransmitters and hormones are both chemicals released by cells, which carry information to other cells in the body. ✓ Neurotransmitters are produced by presynaptic neurones and hormones are secreted by cells in endocrine glands. ✓ Neurotransmitters just diffuse across synaptic clefts, but hormones are carried all over the body in the blood. ✓ They both bind to receptors in cell surface membranes and make things happen in the cell. ✓ Hormones sometimes go right into the cell and act as transcription regulators, but neurotransmitters don't do this. ✓

🖉 This is a well constructed answer, based largely on the candidate's own knowledge rather than being taken from the article. Each sentence includes information about neurotransmitters and hormones, so the examiner can clearly see that a comparison is being made. The candidate has described three differences and two similarities. 5/5

(ii) They may sense the blue light, using cryptochromes or phototropins. ✓ The greater the light intensity, the more these photoreceptors will be stimulated. Or they might use phytochromes, which are sensitive to red and far-red light. P_r gets changed to P_{fr} in the daytime and the other way round in the dark. ✓

🖉 Credit has been given for mention of blue-light receptors. More could have been said about how P_r and P_{fr} are involved in sensing day length, but this is just enough for 2 marks. 2/2

(iii) Most organisms that have been looked at seem to have in-built clocks. Very different organisms all have similarities in the way their clocks work, and this must be determined by genes. ✓ However, the clocks are also affected by the length of light-dark cycles, so they are fine-tuned by the environment. ✓ For example, in humans the body temperature is lower at night than in the daytime, and this is controlled by genes, ✓ because it happens even if we are in constant light or constant dark. ✓ But when students lived in constant light it was found their clocks ran on a cycle longer than 24 hours, so this shows the basic genetic clock is being kept on a 24-hour cycle by being exposed to light and dark. ✓ This isn't the same for all animals. For example ptarmigan (birds) kept in constant dark or constant light (in the Arctic) didn't show any rhythm in their activity, and the same thing happened with reindeer that were used to living in the Arctic. ✓ So their clocks must be genetically different ✓ from human clocks.

This candidate has done really well with this difficult question. He or she has found different pieces of evidence showing that the internal clock is basically governed by genes (in the second sentence and the last one) and also that it is affected by environment (for example what happens to the clock in the students in constant light and the ptarmigan and reindeer in constant light or dark). There are numerous other examples in the article that could have been used, but this is enough for full marks. 7/7

Sample paper

Question 1

The thyroid hormone T4 can act as a transcription factor. An investigation was carried out into the effects of T4 on the expression of three genes in the cornea of developing chick embryos.

Corneas of nine-day-old chick embryos were either left untreated or injected with 2.5 µg of T4. The quantities of mRNA transcribed from four different genes, DIO3, THRB, OGN and CHST1, were measured over the next four days. These quantities were compared against the quantity of mRNA transcribed from a gene called GAPD, which is expressed uniformly throughout this period of chick development and is known not to be affected by T4.

The diagram shows the results.

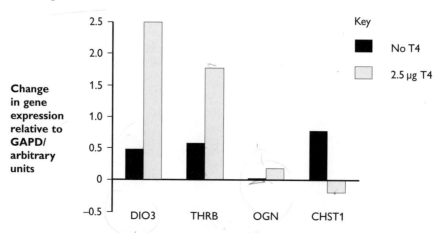

(a) Suggest why the expression of the genes was compared against that of GAPD, rather than being recorded as a straightforward quantity of mRNA produced. (2 marks)

(b) (i) Describe the effects of T4 on the expression of the four genes. (3 marks)

(ii) Explain how T4 could have these effects. (3 marks)

Total: 8 marks

■ ■ ■

Candidate A

(a) So that it was a fair test.

📝 This is not enough for a mark. 0/2

(b) (i) T4 increases the expression of DIO3, THRB and OGN ✓ and decreases it for CHST1. ✓

📝 Correct, but more is needed for a third mark. 2/3

(ii) It could be a transcription factor. ✓ It could bind to the DNA where a gene starts ✓ and stop enzymes binding with it so it won't be used.

📝 Correct, but more detail could be provided about the effect of the transcription factor on the gene. 2/3

Candidate B

(a) This is in case all the genes are expressed more and more mRNA is made as the chick embryo gets older and gets bigger. ✓ We know GAPD is expressed just the same all the time, so this gives us a sort of fixed point ✓ we can compare the other genes against.

📝 Correct. 2/2

(b) (i) T4 increases the expression of DIO3, THRB and OGN. ✓ For DIO3 it goes up from just below 0.5 to 2.5, and for THRB it goes from just above 0.5 to 1.8, so the effect is larger for DIO3. ✓ T4 decreases the expression of CGST1. ✓

📝 Three clear and relevant points made. 3/3

(ii) Some hormones act as transcription factors, ✓ or sometimes they activate transcription factors in the cell. The T4 could go into the cell through the cell membrane and into the nucleus and bind with the DNA ✓ at a particular point. This could stop the enzymes binding with it ✓ that are needed for transcription, ✓ so the gene would not be transcribed into mRNA and would not be expressed. Or it might make it easier for the enzymes to bind, ✓ so it would be expressed more than usual.

📝 A clear and full answer. 3/3

Question 2

The diagram shows a relaxed sarcomere in a muscle fibre.

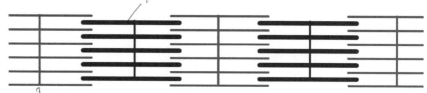

(a) On the diagram, label:

   (i) a myosin filament   (1 mark)

   (ii) a Z line   (1 mark)

(b) Draw a simple diagram to show how this sarcomere would appear when the muscle is contracted.   (3 marks)

(c) Explain how the arrival of an action potential can cause myosin filaments in a sarcomere to form temporary bonds with actin filaments.   (5 marks)

(d) Describe the role of ATP in muscle contraction.   (2 marks)

                                                   Total: 12 marks

■ ■ ■

Candidate A

(a) (i) and (ii)

Z line ✗   Myosin filament ✓

One correct and one incorrect. 1/2

(b)

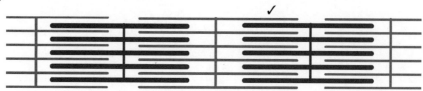

✓

The candidate has correctly shown the new positions of the actin and myosin filaments, but no bridges are shown and the sarcomere is still the same length as in the original diagram. The actin filaments are incorrectly shown as getting longer. 1/3

(c) The action potential goes along the muscle cell membrane ✓ and this makes calcium ions ✓ go into the sarcomere. This makes troponin and tropomyosin move ✓ so the myosin can bind with the actin and pull it along.

   ✒ All correct, but more needed for full marks. 3/5

(d) ATP provides the energy to make the actin filaments move in between the myosin filaments.

   ✒ It is true that ATP provides the energy for muscle contraction, but we need to know exactly what it does. 0/2

Candidate B

(a) (i) and **(ii)**

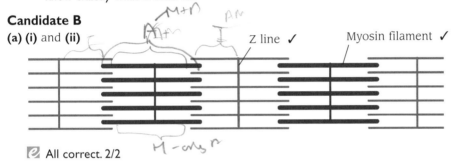

Z line ✓   Myosin filament ✓

   ✒ All correct. 2/2

(b)

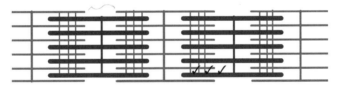

   ✒ Entirely correct. 3/3

(c) The action potential sweeps along the sarcoplasmic reticulum. ✓ This makes Ca²⁺ channels open, ✓ so calcium ions go down their concentration gradient out of the cisternae in the sarcoplasmic reticulum ✓ and go in amongst the myosin and actin filaments. They make the troponin and tropomyosin on the actin filaments move ✓ out of the way so the myosin heads can bind with them, forming actomyosin bridges.

   ✒ A good answer. For full marks, the candidate should have explained that the troponin and tropomyosin cover the binding sites on the actin filaments. 4/5

(d) ATP binds to the myosin heads ✓ when they have moved the actin along. The heads contain an ATPase, and this hydrolyses the ATP into ADP and Pᵢ, which provides energy ✓ to make the myosin let go ✓ of the actin and flip back to its normal position.

   ✒ A complete and correct answer. 2/2

Question 3

The flow diagram shows part of the metabolic pathway of glycolysis.

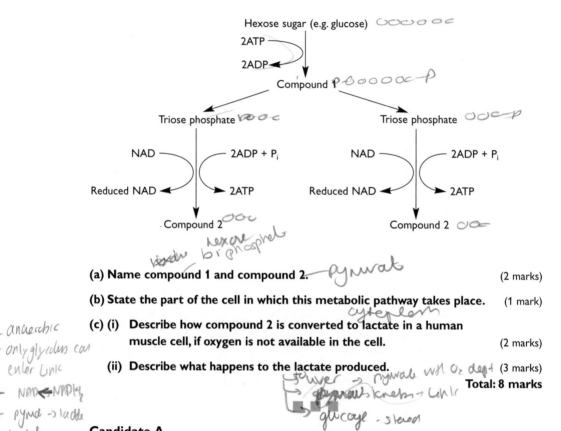

(The diagram shows, with handwritten annotations:)

Hexose sugar (e.g. glucose) ~~~~~~ c

2ATP
2ADP

Compound P ~~~~~~ P

Triose phosphate ~~ o c — Triose phosphate ~~~ p

NAD — 2ADP + P$_i$ NAD — 2ADP + P$_i$

Reduced NAD ← 2ATP   Reduced NAD ← 2ATP

Compound 2 ~~~   Compound 2 ~~

(handwritten:) hexose bi phosphel

(a) Name compound 1 and compound 2. *(handwritten: pyruvat)*   (2 marks)

(b) State the part of the cell in which this metabolic pathway takes place. *(handwritten: cyteplasm)* (1 mark)

(c) (i) Describe how compound 2 is converted to lactate in a human
muscle cell, if oxygen is not available in the cell.   (2 marks)

(handwritten left margin:)
- anaerobic
- only glycolys can enter Link
- NAD ← NADH₂
- pyruv → lactic
- lactal dehydrogenor

(ii) Describe what happens to the lactate produced.   *(handwritten: liver with O₂ dept)* (3 marks)

Total: 8 marks

(handwritten:) liver → pyruvate with O₂ dept / glucose krebs + Link
glucose - stored

Candidate A

(a) Compound 1 is fructose phosphate. ✗ Compound 2 is pyruvate. ✓

> Compound 1 should be bisphosphate, not phosphate (it has two phosphate groups attached to it.) 1/2

(b) Cytoplasm. ✓

> Correct. 1/1

(c) (i) This is anaerobic respiration. The pyruvate is changed into lactate so it doesn't stop glycolysis happening.

> The candidate has not answered the question. 0/2

(ii) It goes to the liver, which breaks it down. ✓ This needs oxygen, which is why you breathe faster than usual when you've done a lot of exercise.

🖉 Again, much more is needed for full marks. The comment about breathing rate is correct, but is not relevant to this particular question. 1/3

Candidate B

(a) Compound 1 is hexose bisphosphate. ✓ Compound 2 is pyruvic acid. ✓

🖉 Both correct. 2/2

(b) Cytoplasm. ✓

🖉 Correct. 1/1

(c) (i) It is combined with reduced NAD, ✓ which is oxidised back to ordinary NAD. The enzyme lactate dehydrogenase ✓ makes this happen.

🖉 Entirely correct. 2/2

(ii) The lactate diffuses into the blood and is carried to the liver cells. ✓ They turn it back into pyruvate again, ✓ so if there is oxygen it can go into a mitochondrion and go through the Krebs cycle. Or the liver can turn it into glucose again, ✓ and maybe store it as glycogen.

🖉 All correct and relevant. 3/3

- lactd → in blod   → liver cells
- lactds → pyruval
- with O₂ fra O₂ dept pyuve → nttached → Krbs
- or glucose
- or stored as glycogen.

Question 4

Thale cress, *Arabidopsis*, has several different receptors for blue light. These include two types of protein, cryptochromes and phototropins.

A study was carried out into the functions of two cryptochromes, cry1 and cry2, and two phototropins, phot1 and phot2.

Seedlings with mutations in the genes for different combinations of these four photoreceptors were exposed to unilateral blue light. The degree of curvature of the shoot towards the light was measured. Some of the results are shown in the graph.

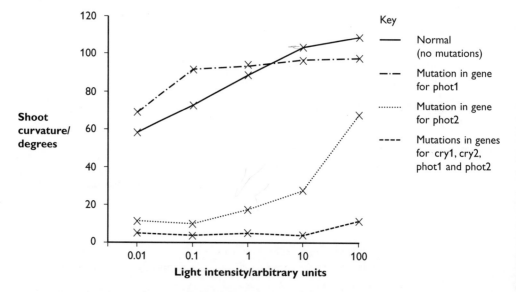

(a) Using the information in the graph, describe the response of normal seedlings to unidirectional blue light. (2 marks)

(b) Explain what the graph suggests about the involvement of each of the following photoreceptors in phototropism.

   (i) phot1 (2 marks)

   (ii) phot2 (2 marks)

   (iii) cry1 and cry2 (2 marks)

(c) Explain how IAA (auxin) is involved in the phototropic response of plant shoots. (4 marks)

Total: 12 marks

■ ■ ■

Candidate A

(a) The greater the light intensity, the more they bend towards the light. ✓

> 🖉 A correct general description, but no detail so no second mark. 1/2

(b) (i) When the plants did not have any phot1, they bent towards the light almost as much as the normal ones, ✓ so phot1 can't be involved. ✓

> 🖉 Correct. 2/2

(ii) The ones with no phot2 only bent towards the light when the intensity was high, ✓ so this must be involved in bending towards the light.

> 🖉 Correct, but more needed for the second mark. 1/2

(iii) The plants without cry1, cry2, phot1 or phot2 hardly bent towards the light at all, ✓ so cry1 or cry2 must be really important. ✓

> 🖉 Correct. 2/2

(c) Auxin collects on the shady side of the shoot ✓ and makes it grow more than the other side, ✓ so the shoot bends towards the light.

> 🖉 There is just enough here for two marks, but this answer is missing a lot of detail. 2/4

Candidate B

(a) The seedlings bend towards the light. When the light intensity is increased from 0.01 to 100 arbitrary units, the amount of bending goes up ✓ from about 70 to 100, an increase of 130 degrees of curvature. ✓

> 🖉 A good answer, including some manipulation of the data shown on the graph. 2/2

(b) (i) The graph shows that the seedlings with a mutation in the gene for phot1 showed a similar response to those that had normal phot1, ✓ so phot 1 cannot be important in this response. ✓

> 🖉 Correct. 2/2

(ii) The bending without phot2 is much less than in the normal seedlings, so phot2 must be important. ✓ But at high light intensities there is still quite a lot of bending, so it cannot be the only photoreceptor involved. ✓

> 🖉 An excellent answer. 2/2

(iii) When the seedlings didn't have cry1, cry2, phot1 or phot2 they hardly bent at all apart from a little bit at very high light intensities, so at least one of cry1 or cry2 must be really important ✓ for phototropism because they didn't bend as much without them as when they are just lacking phot1 or phot2. ✓

> 🖉 Again, a very good answer. 2/2

question

neuble

(c) IAA is made in the growing point at the tip of the shoot and then moves down the shoot. ✓ When blue-light receptors ✓ pick up unidirectional light, they make the IAA accumulate on the shady side. ✓ The IAA switches on genes ✓ for expansins, ✓ which make cell walls stretchy so the cells can get longer. ✓ This makes the cells on the shady side get longer than the cells on the light side, ✓ so the shoot bends towards the light.

✍ A full and clearly explained answer, with plenty of relevant detail. 4/4

Question 5

In the **1960s,** many athletes in some eastern European countries were regularly given anabolic steroid hormones. **The graph below shows how the performance of a woman shot putter changed during a 20-week training period. The training she did during this period was identical to her training before she began taking the steroid hormones.**

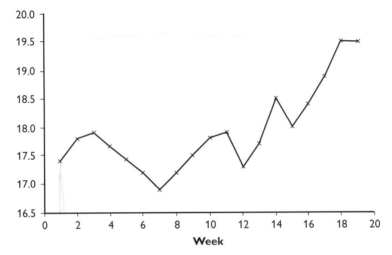

(a) **Calculate the percentage improvement in the woman's performance over the 20 week period. Show your working.** (3 marks)

$\frac{2.\cancel{2}1}{19.5} \times 100 = 12.1\%$

$19.5 - 17.\cancel{4} = 2.25$

(b) **The researchers concluded that the improvement in performance was caused by the steroid hormone. Discuss the validity of this conclusion.** (3 marks)

(c) **The hormone that the woman took was similar to testosterone.**

Discuss the acceptablity of the use of such hormones to enhance athletic performance. (5 marks)

Total: 11 marks

■ ■ ■

Candidate A

(a) 19.5 – 17.4 = 2.1 increase ✓

2.1/19.5 × 100 = 10.76923% ✗

> The increase has been correctly calculated, but it should then have been divided by the original distance, not the final one, to calculate the percentage change. (It is also not correct to give an answer to so many decimal places.) 1/3

(b) The person did the same training all the time, so it can't have been caused by differences in the training. ✓ But maybe the training was making her get better anyway. We don't know whether she was getting better before she started taking the steroids, ✓ so maybe it was just the training making her get better. And she got worse in the first 7 weeks.

> 🖉 The candidate is suggesting that perhaps this improvement is just a continuation of a similar pattern that was happening before the shot putter began taking steroids, and this is a valid criticism. 2/3

(c) The testosterone might make her more male, so it might make her ill and maybe mess up her reproductive system. ✓ It might make her stronger with bigger muscles, ✓ so that would give her an unfair advantage over women who didn't take the hormone. ✓

> 🖉 This is all fair comment, but the question asked the candidate to 'discuss' the issue, so we need to see arguments on the other side as well as arguments against using the hormone. 3/5

Candidate B

(a) 19.5 – 17.4 = 2.1 increase ✓

$2.1 \div 17.4 \times 100 = 12.1\%$ ✓

> 🖉 Correct, and the answer is given to one decimal place, to match the figures read off from the graph. 3/3

said it couldn't be

(b) We are told that the training regime during this trial was the same as before it, so it cannot be a change in the training that caused the increase. ✓ So it could be the hormone, but we can't be certain of that. The experiment only used one person, so it could just be chance ✓ or something else we don't know about that made her improve so much. We need to do a trial with lots of women and really control all the other variables, ✓ like what they eat and how old they are and how motivated they are to improve their performance.

> 🖉 The candidate has correcly pointed out that it is possible another uncontrolled variable was involved in the shot putter's performance, and also that we can't really conclude anything from just one person's results. 3/3

absch tut

(c) Some people think that all athletes should be able to take whatever they want to improve their performance, and if everyone could do this then it would be fair ✓ and you wouldn't get things happening like when someone takes a food supplement or a medicine and then gets disqualified ✓ because there was something in it they didn't know about. ✓ And anyway we can't detect all the drugs, so someone might be taking them and we can't tell. ✓ But most people think it is best if athletes don't take powerful drugs like this because it can damage their health ✓ and it makes the competition unfair. ✓

> 🖉 Some interesting points made, and both sides of the argument addressed. 5/5

Question 6

(a) **Outline the role of the hypothalamus in thermoregulation in humans.**   (3 marks)

(b) **Describe how effectors help to lower core body temperature when it rises too high.**   (6 marks)

Total: 9 marks

Candidate A

(a) The hypothalamus detects changes in body temperature and sends messages to the skin to make it do things to make the body temperature normal again. This is negative feedback.

  There is nothing quite good enough in this answer to get any marks. 0/3

(b) The capillaries in the skin get wider so there is more blood in them ✓ so it gets colder. And the sweat glands make more sweat which cools the skin. ✓

  The candidate again has not given enough detail to pick up many marks. 2/6

Candidate B

(a) The hypothalamus contains thermoreceptors ✓ which sense the temperature of the blood. ✓ It also receives nerve impulses from temperature receptors in the skin, ✓ which tell it about the temperature of the environment. If either of these inputs suggests that the core temperature of the body might be too high, or going to get too high, then the hypothalamus sends action potentials ✓ to effectors ✓ in the skin to take actions that will bring it down. And the opposite if it is going too low.

  A very full answer, mentioning the inputs to the hypothalamus and how it responds to these. 3/3

(b) The sweat glands produce extra sweat, ✓ which goes up the sweat ducts and onto the skin surface. The water in it evaporates. ✓ Water has a high latent heat of evaporation, which means it needs a lot of energy to do it, so it takes heat energy from the skin and makes it feel cooler. ✓ Also, the smooth muscle in the walls of the arterioles ✓ supplying the capillaries near the skin surface relaxes, ✓ so the arterioles dilate ✓ and more blood flows to the surface and loses heat by radiation to the air. ✓

  Plenty of relevant detail, so full marks. 6/6

Question 7

The scientific document you have studied (pages 55–60) is adapted from articles on refractory period and the cerebral cortex in *Biological Sciences Review*. Use the information in the document and your own knowledge to answer the following questions.

(a) The cell surface membranes of axons have many different methods of controlling the passage of ions through them, including sodium-potassium pumps and voltage-gated cation channels.

Explain how the sodium-potassium pump differs from a voltage-gated cation channel. (2 marks)

(b) The diagram shows an action potential.

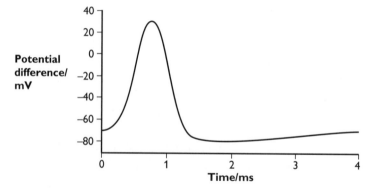

(i) Indicate on the diagram when the voltage-gated sodium ion channels open. (1 mark)

(ii) Indicate on the diagram when the voltage-gated potassium ion channels open. (1 mark)

(iii) Explain what is meant by positive feedback, and describe its importance in the generation of an action potential. (2 marks)

(c) (i) Explain the difference between absolute refractory period and relative refractory period. (2 marks)

(ii) Describe how the existence of a refractory period enables different organs in the body to function correctly. (6 marks)

(d) (i) What is the cerebral cortex? (3 marks)

(ii) Outline the principles of fMRI. (3 marks)

(iii) Explain how a cortical map can be produced using fMRI. (3 marks)

(e) Habituation is a simple type of learning. Discuss the extent to which learning to play a musical instrument differs from habituation. (7 marks)

Total: 30 marks

■ ■ ■

Candidate A

(a) The sodium-potassium pump moves ions against their concentration gradients, but a voltage-gated channel just lets them through down their concentration gradient. ✓

📝 This is correct, but more facts are needed. 1/2

(b) (i) and **(ii)**

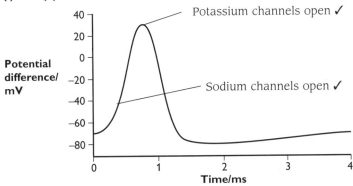

📝 Both correct. 2/2

(iii) This is when something goes up and then it is made to go up even more. ✓ When an action potential starts up, it goes higher and higher.

📝 There is just enough here for one mark, but the comment about the action potential does not give any correct information. 1/2

(c) (i) An absolute refractory period is the time after an action potential has happened when the neurone can't do another one no matter what happens. ✓ A relative refractory period is when it can do another one if there is a really strong depolarisation. ✓

📝 This is not well worded, but the candidate does convey some correct information about these two types of refractory period, using information from the article. 2/2

(ii) If there wasn't a refractory period, then you would just get fibrillation ✓ in the heart because there wouldn't be time for the muscle to relax in between contracting. ✓ It's also important in the ear so that you can tell how loud a sound is ✓ by there being more action potentials close together when it is loud than when it is quiet. If there wasn't a refractory period then they would all be the same time apart all the time. ✓

📝 Again, the candidate has struggled to give a clear explanation, but he or she has found two good examples in the article and has made a reasonable attempt at using them to answer the question. 4/6

(d) (i) It is the outside part of the cerebrum. ✓ It is very folded. ✓ It receives information from our senses and coordinates them so we make decisions about doing the right actions. ✓

📝 This is enough to get all the marks. 3/3

(ii) fMRI uses a big magnet that you lie inside. ✓ It measures the magnetic field ✓ in your brain, and this is different depending on how much oxyhaemoglobin ✓ there is. A lot of oxyhaemoglobin means there is a lot of activity in that part of the brain.

📝 Again, a good answer. 3/3

(iii) You get a person to lie in a fMRI magnet and tell them to do different things ✓ and see which parts of the cortex are most active. ✓

📝 Quite a good answer, which makes two clear and relevant points. 2/3

(e) Habituation is when you get used to something so you stop reacting to it. ✓ Even really simple things like snails and sea slugs have habituation, ✓ but they can't play musical instruments. Habituation just involves one simple nerve pathway, but playing a musical instruments uses lots of different ones, and different parts of the brain. Habituation may not involve the cortex at all because it isn't something we think about. ✓ But playing an instrument uses many different areas in the cortex ✓ and the parts of it that receive lots of stimuli — like from the fingers of the left hand if you play a violin — get bigger. ✓

📝 This is a very difficult question, but this candidate has made a good attempt at it, using his or her own knowledge about habituation, and information from the article about learning to play a musical instrument. 5/7

Candidate B

(a) The sodium-potassium pump is a group of proteins in a cell membrane that moves 2 potassium ions in for every 3 sodium ions out. They are moved against their concentration gradients, using ATP. Voltage-gated cation channels only open when the voltage across the membrane is a particular value, ✓ and they only let one ✓ sort of ion through, not two. The ions move through the channels down their concentration gradients, ✓ not up them.

📝 An excellent answer, making three clear comparative points. 2/2

(b) (i) and **(ii)**

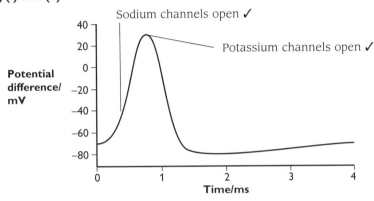

 All correct. 2/2

(iii) In positive feedback, if something changes in one direction then the body responds by making it change even more in the same direction. ✓ For example, when a nerve cell membrane is depolarised, this opens voltage-gated sodium channels ✓ which let the sodium ions flood in so it depolarises even more. ✓ This is what makes the voltage across the membrane shoot up so it becomes positive inside, ✓ making an action potential.

 A good answer. The candidate has used information about positive feedback in this stage of the generation of an action potential in the article, and has put it into his or her own words. 2/2

(c) (i) During an absolute refractory period, no action potential can be triggered at all, ✓ but in the relative refractory period you can trigger an action potential but only with a larger depolarisation than usual. ✓

 All correct. 2/2

(ii) The refractory period makes sure you don't just get a string of action potentials all overlapping each other. ✓ There are small gaps between them. The size of these gaps can be varied, so if there are a lot of action potentials close together in a neurone from a sense organ like the ear, ✓ this means the stimulus is very strong. ✓ You couldn't use this way of telling how strong a stimulus is if there wasn't a refractory period.

Also, in the heart, the refractory period helps to keep all the heart muscle contracting at the right time. ✓ If there isn't a refractory period then it just contracts and relaxes really quickly so you get fibrillation ✓ which means the muscle is just fluttering instead of pumping the blood.

Refractory periods also mean the impulses go in just one direction ✓ — they can't go back where they just came from.

Three different examples are given, each using information from the article but explained in the candidate's own words. 6/6

(d) (i) The cerebral cortex is the outer part of the cortex. ✓ It is highly folded, ✓ and just a few millimetres thick.

Correct, but needs a little more for full marks. 2/3

(ii) fMRI stands for functional magnetic resonance imaging. ✓ It measures tiny differences in magnetic field ✓ in different parts of the brain, which are affected by the amount of oxygenated blood ✓ flowing to them.

Clear and correct. 3/3

(iii) A person lies in a huge cylindrical magnet. They carry out different tasks involving their brain, ✓ or are given different stimuli. ✓ The image from the fMRI show which parts of the cortex are most active ✓ with a particular task or a particular stimulus.

All correct. 3/3

(e) Habituation can be defined as a decrease in the intensity of a response when the same stimulus is given repeatedly. ✓ This doesn't happen when learning to play an instrument. In fact, it is almost the opposite — when you see a particular note on a music score you actually learn to respond more quickly ✓ and strongly to it by playing the right note on the instrument. Also, habituation is an involuntary thing ✓ but learning to play an instrument involves you thinking about it and using your cerebral cortex. ✓ Even really simple animals become habituated to stimuli, whereas learning to play an instrument has so far only ever been done by humans. But there are some similarities, in that in both cases repeated exposure to a stimulus causes a change ✓ in the way that you react to it, so it must involve something happening to synapses ✓ in the brain or other parts of the nervous system.

A really good attempt at a very difficult question. The candidate has succeeded in picking out relevant points from the article, and integrating them with what he or she knows about habituation. 6/7